The Artificial Intelligence Experience:
An Introduction

Copyright© 1985 by Digital Equipment Corporation

Digital Equipment Corporation believes the information in this publication is accurate as of its publication date; such information is subject to change without notice. Digital is not responsible for any inadvertent errors.

The following are trademarks of Digital Equipment Corporation:

the Digital logo, DEC, DATATRIEVE, DECOR, DECnet, DECsystem-10, DECSYSTEM-20, DECUS, DECtalk, FMS, GKS, Microvax, Rainbow, PDP-6, PDP-7, PDP-11, TOPS-20, ULTRIX, ULTRIX-32, UNIBUS, VAX, VAX-11/750, 780, VAXcluster, VAXstation II, VAX 8600, VAX DBMS, VAX rdb/VMS, VAX TDMS, VMS, XCON, and XSEL.

†ALPS is a trademark of Automated Language Processing Systems, Inc.; ART and Inference are trademarks of Inference Corporation; DUCK is a trademark of Smart Systems Technology; EPISTLE and IBM are trademarks of International Business Machines Corporation; EXPERT-EASE is a trademark of Jeffrey Perrone & Associates; GCLISP is a trademark of Gold Hill Computers, Inc.; INTELLECT is a trademark of Artificial Intelligence Corporation; IntelliCorp and KEE are trademarks of IntelliCorp, Inc.; Knowledge Craft and Language Craft are trademarks of Carnegie Group, Inc.; LOGOS is a trademark of Logos Corporation; MACSYMA and Symbolics are trademarks of SYMBOLICS, Inc.; MPROLOG is a trademark of Logicware, Inc.; MS is a trademark of Microsoft Corporation; Prolog II is a trademark of Prologia; Quintus is a trademark of Quintus Computer Systems, Inc.; San Marco is a trademark of San Marco Associates; SMALLTALK-80 and Xerox are trademarks of Xerox Corporation; TIMM is a trademark of General Research Corporation; UNIX is a trademark of AT&T Bell Laboratories; USE.IT is a trademark of Higher Order Software, Inc.; and VAL is a trademark of Unimation, Inc., a Westinghouse company.

®Ada is a registered trademark of the U.S. Department of Defense Ada Joint Program Office; CRAY-1 is a registered trademark of Cray Research, Inc.; CYBER is a registered trademark of Control Data Corporation; MUDMAN is a registered trademark of NL Baroid, NL Industries, Inc.; RAIL and Robovision are registered trademarks of Automatix, Inc.

Susan J. Scown

The Artificial Intelligence Experience:
An Introduction

Contents

	Introduction	1
	Acknowledgments	3
Chapter 1	Artificial Intelligence: Computing Techniques That Help People Solve Complex Problems	7
Chapter 2	Areas of Artificial Intelligence Research and Implementation	15
Chapter 3	The Techniques of Artificial Intelligence	51
Chapter 4	The Languages of Artificial Intelligence	77
Chapter 5	Computer Hardware for Artificial Intelligence	97
Chapter 6	Getting Started with Expert Systems: A Case Study	111
Chapter 7	National Programs and Other Major Cooperative Efforts in AI Research	147
Appendix	Highlights in the Development of Artificial Intelligence	161
	Glossary	169
	Index	175

Introduction

In the past couple of years, many articles about artificial intelligence (AI) have been printed in newspapers, general interest magazines, and computer trade publications. Also in print are technical textbooks on AI and tradebooks that discuss predictions of the impact that AI will have on our lives.

The goal of this book is not to publish new and original information or to speculate about the more exotic possibilities that AI may offer in the future. Its goal is, rather, to synthesize practical information that has become common knowledge in the AI community. In this book you will find the basics, not the most technical information.

At the end of each chapter is an annotated bibliography. The books and articles cited in the bibliographies, supplemented by personal communication with people knowledgeable in AI, both inside and outside of Digital, were the primary sources of information in writing this volume. The annotations are intended to help you decide where to look for additional printed information. The annotations generally highlight only topics in the sources that were discussed in the chapter.

Reading the chapters in the sequence in which they are presented will allow you to utilize ideas presented in earlier chapters in understanding concepts presented in later chapters. Chapter 3, "The Techniques of Artificial Intelligence," is the most technical and difficult chapter. It can be skipped by anyone who does not want an initiation into the "how" of AI.

We hope you will find this volume instructive and thought-provoking. The possibilities that artificial intelligence offers for solving complex problems are very exciting. Digital Equipment Corporation wants to help you put them to work.

May 21, 1985

Acknowledgments

I would like to thank my colleagues at Digital Equipment Corporation for lending their technical expertise, practical assistance, and support in the production of this book. The following list includes technical experts whom I consulted, people who helped in locating printed resources, and writing and editing professionals:

Norma Abel, Judy Bachant, John Barnwell, Art Beane, Pat Bernard, Jerry Boetje, Gary Brown, Ed Bruckhert, Ninamary Buba, Ellie Burns, Ugo Buy, Dick Caruso, Paul Cashman, George Champine, Dave Conroy, Topher Cooper, Tom Cooper, Tom Davis, Mike Dowd, Don Elias, Heinz Franz, Gerhard Friedrich, Al Gauthier, Ken Gilbert, Mike Glantz, Frank Glazer, Bonnie Gutnick, Dan Hall, Paul Hoffmann, Ellen Hurvitz, Bill Kania, Arnold Kraft, Peter Kraus, Lisa Lam, Frank Lynch, Joel Magid, Kent McNaughton, Margaret Meehan, Norio Murakami, Roger Nasr, Dennis O'Connor, Ed Orciuch, Jim Paradise, Steve Polit, Pete Popieniuck, Karl Puder, Neil Pundit, Brian Rees, Hal Shubin, Lee Stommes, Mark Swartwout, Jack Richardson, Christine Tabak, Ellen Teplitz, Allan Titcomb, Walter VanRoggen, Wendy Wilkerson, Bryant York, Bill Yskamp, Dom Zambuto, Steve Zaslaw, and many individuals in the Digital Library Network, notably Marguerite Vierkant, Richard Maxfield, Betsey Cane, and Helen McFadyen.

Many individuals from outside of Digital were also helpful in providing and confirming information and in critiquing this material. Among them were

Diane Ballinger of Automatix, Inc.

Bradley Berg of Smart Systems Technology

Kenny Bergen of NL Baroid, NL Industries, Inc.

Angela Buonocore and George Heidorn of International Business Machines, Inc.

Gerald De Jong of the University of Illinois, Urbana

Ann Drinan of Cognitive Systems, Inc.

Neil Goldman of USC Information Sciences Institute

Richard Jullig and Cordell Green of Kestrel Institute

Janet Kolodner of the Georgia Institute of Technology

Wendy Lehnert of the University of Massachusetts, Amherst

John Lowrance of SRI International

Linda Mayeda and Linda Raneirie of Symbolics, Inc.

Marvin Minsky, Joseph Weizenbaum, Richard Waters, Randall Davis, and Karen Prendergast of the Massachusetts Institute of Technology

Ellen Mohr of Unimation, Inc., a Westinghouse company

Brenda Nashawaty and Clifford Spitz of Artificial Intelligence Corporation

Nils Nilsson and Terry Winograd of Stanford University

Wendy Rauch-Hindin of *Systems and Software*

Roger Schank and Stephen Slade of Yale University

Herbert Simon, John McDermott, Allan Newell, and Charles Forgy of Carnegie-Mellon University

Guy Steele, Jr. of Thinking Machines, Inc.

Ken Thompson of AT&T Bell Labs

David Waltz of Brandeis University

Joe Weaver of Higher Order Software, Inc.

Jennifer Zimmerman of Quintus Computer Systems, Inc.

and many other individuals from companies and organizations including Automated Language Processing Systems, Inc.; Barry Wright Corporation; Bolt Beranek and Newman, Inc.; Carnegie Group, Inc.; Control Data Corporation; Cray Research, Inc.; Franz Inc.; General Research Corporation; Gold Hill Computers, Inc.; LISP Machine, Inc.;

Logos Corporation; Octek, Inc.; Odetics, Inc.; PERQ Systems Corporation; Rand Corporation; Sun Microsystems, Inc.; Teknowledge, Inc.; Tektronix, Inc.; the U.S. Department of Defense Advanced Research Projects Agency; the U.S. National Bureau of Standards; Weidner Communications Corporation; and Xerox Corporation.

A special thank you goes to Walter Hamscher of the Massachusetts Institute of Technology for thoughtful comments on technical topics.

Chapter 1

**Artificial Intelligence:
Computing Techniques That Help
People Solve Complex Problems**

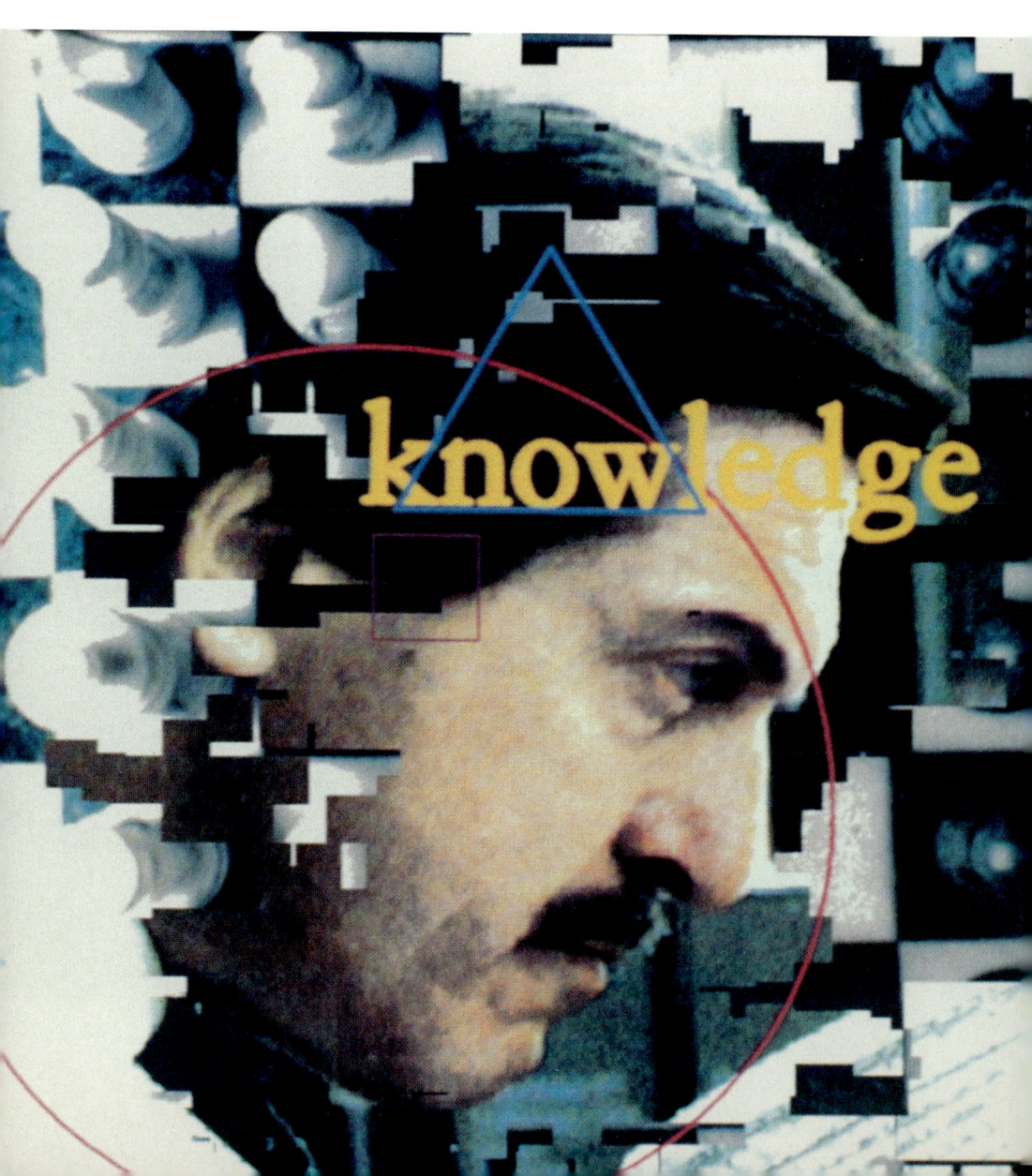

Artificial intelligence is as unsettling a title for a technology as has ever been coined. The phrase seems to challenge our human pride in possessing "real" intelligence. Intelligence means, among other things, the capacity to acquire and use knowledge, to reason with it, and to solve problems effectively with it. But why should we attempt to build these abilities into computers?

There are a number of reasons researchers have worked to develop an artificial variety of intelligence. One reason is to shed light on the human variety of intelligence by attempting to model it with computers. Another is to make computers easier to use by making them operate more like their human users. And with artificial intelligence techniques, many complex problems can be solved that conventional programming methods have not been able to solve efficiently or cannot solve at all.

Now that artificial intelligence (AI) has emerged from the research laboratory, there are additional reasons for interest in this area of computer science. Some business people see AI as providing leverage over their less technologically advanced competitors. To many countries, AI means technologies and commercial opportunities that can improve their security and their competitive positions in the world economy. In many areas, both governmental and private, the ability to capture and utilize knowledge is cited as a key resource for coping with complex real world problems, and AI offers methods for doing this.

How Do We Define AI?

Artificial intelligence is a relatively young field (having existed only for about thirty years) and has, as yet, no strict boundaries or definitions. Some people, particularly in the research community, define AI narrowly to include only techniques that use knowledge to manipulate data in ways that generate novel inferences that were not explicitly programmed. This definition excludes systems that work on the kinds of problems AI addresses and do use techniques common in AI, such as search, but that do not make novel inferences. The difference between the narrow and broad definitions comes into play primarily in discussion of practical applications to which AI has led. For instance, in the expert systems area, some applications demonstrate the ability to make novel inferences but some do not. Broader definitions of AI's turf, particularly in the commercial world, include any applications that use AI techniques, whether or not the particular application can make novel inferences. Some people applying AI techniques call their systems "advanced systems" rather than AI systems in order to reserve the narrower definition to truly "intelligent" systems and to avoid arguments about what AI is.

Here is how some AI experts define artificial intelligence:

- Marvin Minsky, Donner Professor of Science at the Massachusetts Institute of Technology and a founder of the field of AI, has stated that "artificial intelligence is the science of making machines do things that would require intelligence if done by men."[1]
- Patrick H. Winston, director of the Artificial Intelligence Laboratory at the Massachusetts Institute of Technology, writes: "The goals of the field of Artificial Intelligence can be defined as...[attempting] to make computers more useful [and] to understand the principles that make intelligence possible."[2]
- Nils Nilsson, chairman of the Department of Computer Science at Stanford University, says: "The field of Artificial Intelligence has as its main tenet that there are indeed common processes that underlie thinking and perceiving, and furthermore that these processes can be understood and studied scientifically.... In addition, it is completely unimportant to the theory of Artificial Intelligence *who* is doing the thinking or perceiving – man or computer. This is an implementation detail."[3]
- Bruce G. Buchanan, adjunct professor of computer science, research; and Edward A. Feigenbaum, principal investigator of the Heuristic Programming Project; both of Stanford University, write: "Artificial intelligence research is that part of computer science that investigates symbolic, nonalgorithmic reasoning processes and the representation of symbolic knowledge for use in machine intelligence."[4]

For the purposes of this book, AI is a growing set of computer problem-solving techniques that are being developed to imitate human thought or decision-making processes or to produce the same results as those processes. It provides us with a set of labor-saving tools for problems that can be solved by the processing of knowledge.

AI systems can help people multiply their expertise by storing and manipulating knowledge they supply to computer systems. AI also helps solve problems that have multiple solution paths and/or for which there is no known algorithm. In fact, AI systems sometimes help people discover previously unknown algorithms for problems that can then be solved by conventional computing methods. Computers do these things well today when confined to a particular subject area in a task that can be solved by processing a large amount of knowledge.

Artificial intelligence can't solve all programmable problems, but it can certainly solve some well-chosen, difficult problems. With AI, quite a bit of routine work can be handed over to the computer, and human beings can concentrate on problems requiring greater creativity, allowing them to fulfill more of their potentials. And with each advance in the art of AI, the border between routine and creative work can move out further into what was previously the creative territory. In fact it is a commonplace in the AI research world that what was formerly considered AI eventually becomes "just computer science."

A Case in Point

An example of how AI methods are helping people solve difficult problems in the commercial sector is NL Baroid's expert system MUDMAN.® NL Baroid is a $400-million-per-year company whose product is *drilling mud*, a lubricant needed in drilling oil wells. Baroid invented drilling mud in the 1920s and is now the largest mud company in the world.

Sometimes when Baroid sells drilling mud to an oil company, it also sells the services of a mud engineer to stay at the site and solve problems. It takes about three years to train a mud engineer. Baroid has a knowledge base of over 60 years of mud experience, both in written reports and in the knowledge of mud experts with 30 to 40 years in the field. They wanted to make that knowledge available to mud engineers in the field.

When mud engineers call upon personal knowledge to solve a problem, a plausible mechanism to describe this process is that they search through their memories, matching the current situation to a previous pattern. Then they may apply rules of thumb to solve the problem. No conventional algorithms are universally applicable, so no conventional program can reproduce the needed expertise and solve the business problem. A system like MUDMAN, however, uses AI techniques of pattern matching and the application of heuristics, or rules of thumb, to solve the problem the way an expert would, using this model of the thinking process.

The inputs to MUDMAN include the specifications of the type of mud needed in a particular well and the chemical and physical properties of the mud that is actually present. MUDMAN compares the specifications to the actual properties, provides an analysis of drilling problems, and recommends corrective treatments.

MUDMAN was developed in a joint effort by Baroid and Professor John McDermott (acting department head and principal scientist, Department of Computer Science, Carnegie-Mellon University) and associates at Carnegie-Mellon (CMU). The developers at CMU did the feasibility studies and applied research and, during the initial joint development period, trained Baroid personnel in AI techniques. Since CMU turned over the framework for MUDMAN to Baroid in January 1984, Baroid has had full responsibility for field testing, updating, enhancement, and modification.

MUDMAN was specifically designed for sale to Baroid's customers, which are oil companies. Baroid has described MUDMAN as the first expert system sold as a commercial product to the oil industry.

When Is a System Intelligent?

How do AI researchers decide when they have developed a system that successfully exhibits artificial intelligence? That depends on their working definition of AI and on the task. Systems whose purpose is to play a popular game like checkers or chess have been evaluated by performance ratings on the same scale on which we rate masters of the game. Some AI systems have been rated as masters of checkers; AT&T Bell Laboratories reports that its chess-playing system, Belle, has been rated at the master level.

A classic, though not rigorous, test of the intelligence of a system was formulated by the British computer scientist Alan Turing. Turing's Test requires two people, a computer, a dividing wall, and a communication device like a teletype, by which both a human and a machine can communicate on an equal basis. One person, the interrogator, is positioned on one side of the wall, and the computer and the other person are positioned on the other side of the wall. The interrogator communicates with the computer and the other person by means of the communication device, asking them both questions and trying to determine which is the machine. If the computer fools the interrogator, Turing said, we can conclude that it "thinks." No strong claims have been made that a computer has passed Turing's Test in any but highly restricted formats. This is due to the computer's inability to interpret subtle nuances of meaning, read between the lines, and interpret motivations of the speaker, as well as to the computer's lack of reference to the wide range of subject matter typical of adult thought.

To evaluate an expert system like PUFF, which diagnoses lung diseases, we can compare the system's solutions to those of human experts. PUFF's diagnoses were demonstrated to have a 96 percent rate of agreement with the physician whose knowledge was used to program the system and an 89 percent rate of agreement with an independent physician. The rate of agreement between the physicians themselves was 92 percent. PUFF's rate of agreement with the experts qualifies the system as performing quite "intelligently."

Even with artificial intelligence methods, computers can't match human beings in situations that require rapid processing of sense information like seeing, hearing, or tactile sensing or when it comes to using "common sense" (knowledge of the world that helps people determine the correct interpretation, significance, and reasonable implications of statements in varying contexts). Computers also have a long way to go before they can communicate with human beings with an expert's depth of understanding of a particular field. These limitations arise in large part from the computer's lack of ability to blend information from different realms together in an integrated way. Nevertheless, as we will see in later chapters, computers can serve as efficient, intelligent assistants to their human managers.

While research remains a very important part of artificial intelligence, practical applications are already in use today. In fact, the potential return on investment is so encouraging that many large corporations have created inhouse AI research groups. Some of the main areas of research and implementation at the moment are expert systems, "natural language" systems, robotics and sensing systems, and automatic programming. Because expert systems is the area of AI technology that has achieved the most success in terms of practical applications, this book will devote more attention to the expert systems applications area than to the others.

In this book, we will also look at the basic concepts of artificial intelligence, the hardware and languages used to develop AI programs, how you can start using AI technology, and major research efforts. This will give you an overview of what artificial intelligence can do now and what researchers expect the technology to accomplish in the future.

Annotated Bibliography

Alexander, Tom, "Teaching Computers the Art of Reason," *Fortune,* May 17, 1982, 82-92. This overview of AI research is written in terms understandable to the layman. It is the first article in a three-part series.

Alexander, Tom, "Practical Uses for a Useless Science," *Fortune,* May 31, 1982, 138-145. This article surveys AI applications.

Alexander, Tom, "Computers on the Road to Self-Improvement," *Fortune,* June 14, 1982, 148-160. This article discusses languages, hardware, and future developments expected in AI.

"Artificial Intelligence Is Here," *BusinessWeek,* July 9, 1984, 54-62. This article discusses applications, commercial ventures, and future directions in AI research.

Boden, Margaret A., *Artificial Intelligence and Natural Man,* New York, Basic Books, Inc., 1977. Boden's book describes artificial intelligence in a way that stresses its human relevance, particularly in relation to psychology and philosophy. Boden uses plain language and makes the workings of natural language programs clear.

McCorduck, Pamela, *Machines Who Think,* San Francisco, W. H. Freeman and Company, 1979. McCorduck's book provides a very readable history of the technological developments and the theoretical currents that led to the emergence of the field of AI.

Nilsson, Nils J., "Artificial Intelligence Prepares for 2001," *The AI Magazine,* Winter 1983, 7-14. Nilsson's article discusses recent achievements of and investigations in AI, and provides a viewpoint on the sphere of inquiry proper to AI. Nilsson offers challenges to researchers in the field.

Schank, Roger C., "The Current State of AI: One Man's Opinion," *The AI Magazine*, Winter/Spring 1983, 3-8. Schank discusses the sphere of inquiry he believes is proper to AI.

Notes

1. Boden, Margaret A., *Artificial Intelligence and Natural Man*, New York, Basic Books, Inc., 1977, p. 4.

2. Winston, Patrick H., *Artificial Intelligence* (2nd ed.), Reading, Massachusetts, Addison-Wesley Publishing Company, 1984, p. 1.

3. Nilsson, Nils, J., "Artificial Intelligence," Technical Note 89, Menlo Park, California, Artificial Intelligence Center, Stanford Research Institute, April 1974, p.1.

4. Davis, Randall, and Douglas B. Lenat, *Knowledge-Based Systems in Artificial Intelligence*, New York, McGraw-Hill International Book Company, 1982, p. xv.

Chapter 2

**Areas of Artificial Intelligence
Research and Implementation**

Researchers have set goals for artificial intelligence technology to achieve during the 1990s and later, but AI can also claim achievements here and now. Successful applications of the technologies have demonstrated that artificial intelligence methods can solve massively complicated problems.

For example, AI systems have already been developed that

- Allow the user to type queries to a database in everyday language, rather than in a programming language.
- Recognize objects in commonplace scenes with machine vision.
- Generate recognizable humanlike speech from computerized text.
- Recognize and interpret a small vocabulary of continuous human speech.
- Solve problems utilizing the encoded knowledge of experts in a variety of fields.

Several countries, most notably the United States, Japan, the United Kingdom, and those of the European Economic Community, are sponsoring artificial intelligence research (see Chapter 7). Many large companies are also investing in artificial intelligence technology, investigating, both on their own and in cooperative ventures with universities and other companies, how AI can be applied to solve real problems and save money.

Current applications of AI technology fall primarily into four fields—expert systems; natural language systems; robotics, vision, and sensing systems; and automatic programming. This chapter will give you an overview of what artificial intelligence offers in these four fields at the present.

Expert Systems

One of the most economically rewarding areas of artificial intelligence implementation today is that of expert systems, also called knowledge-based systems. An expert system applies AI reasoning and problem-solving techniques to knowledge encoded about a specific problem domain in order to simulate the application of human expertise. The effectiveness of expert systems comes from the amount of knowledge provided for them. No high-powered, generic reasoning methods have been found that can create intelligent behavior without a sufficient fund of knowledge to reason with. However, some very simple inferencing methods can get impressive results when they are applied to an adequate base of knowledge.

Expert systems are most often used as intelligent assistants or consultants to human users. They can be used to solve routine problems, freeing the expert for more novel and interesting ones. Expert systems can also bring expertise to locations where a human expert is unavailable or to situations in which an expert's services would be very expensive. Some corporations even see expert systems as a way to collect and preserve a "corporate memory" because an expert system never retires, becomes sick, or leaves.

Among other achievements, expert systems already have

- Acted as trouble-shooting advisers for oil-well drilling equipment.
- Advised physicians on treatments for suspected meningitis and other bacterial infections in the blood.
- Deduced the location of a large molybdenum deposit.
- Configured complicated computer systems in a fraction of the time required by an experienced engineer.

Unlike conventional computer programs, expert systems can tackle problems that require judgmental decisions, the kind people work with every day. For example, the expert system MYCIN can recommend treatments for suspected meningitis and other bacterial infections of the blood by analyzing a physician's observations of the patient. Expert systems can do this sort of humanlike reasoning because they are programmed with the knowledge of experts.

Some expert systems, like some traditional systems, provide their answers in terms of confidence measures, having propagated through the program degrees of certainty associated with pieces of information. Some systems also have explanation facilities that tell how they reached their answers. In some, this merely amounts to providing a list of the steps the system followed during processing. However, some systems also are programmed to give the reason for choosing one alternative rather than another. Without explanation facilities, a user might be reluctant to act on a system's conclusions. For this reason, explanation facilities can be vital to an expert system's usefulness.

In the pioneering days of commercial expert systems, it took many developer-years to bring an expert system up to speed. The time has been reduced as developers have gained familiarity with methods of producing expert systems and as software and hardware that facilitate the process have developed further. While it requires an investment to develop and maintain an expert system, that cost can be quite low compared to the cost of locating and hiring experts (who tend to be expensive and scarce) or trying to develop your own experts through training and experience (which could take years). And the cost of doing without expertise in a vital problem area is all too obvious.

- *Expert Systems Basics*

Artificial intelligence reasoning and problem-solving techniques allow expert systems to draw conclusions that were not explicitly programmed into them. They contrast with traditional data processing techniques, which demand input that is certain, use primarily numeric computation in well-understood algorithms and, when done correctly, produce correct answers. Expert systems, on the other hand, use information that is not always entirely consistent or complete, manipulate it by symbolic reasoning methods without following a numeric model, and still produce satisfactory answers and useful approximations. Naturally, the more complete and correct the knowledge is, the better the system's output is.

Some of the techniques and elements that make possible an expert system's novel inferences are *knowledge acquisition, heuristics, knowledge representation methods*, and *inference engines*.

Knowledge acquisition is the process of extracting and formalizing the knowledge of an expert for use by an expert system. Examples of knowledge are descriptions of objects, identifications of relationships, and explanations of procedures. Software engineers called knowledge engineers specialize in the techniques of knowledge acquisition. They are trained to help experts articulate their experience and rules of thumb. They also decide on the best way to structure this knowledge for use by the system.

Heuristics are rules of thumb an expert has learned or discovered concerning a particular problem area. When encoded in an expert system, these rules help guide the processing through masses of data. Heuristics make the solution process more efficient than blindly searching (see Chapter 3 for a discussion of search methods) through a field of possible solutions.

A *knowledge representation* is a formalized structure and set of operations that embodies the descriptions, relationships, and procedures an expert provides for an expert system. The knowledge represented in a particular program is called its *knowledge base*. Examples of knowledge representation forms include production ("If...then....") rules, frames, and semantic nets. (See Chapter 3 for more information.) Different representations are suited to different types of knowledge and different uses of knowledge. "If...then...." rules lend themselves to the representation of deductive knowledge—situation/action, premise/conclusion, antecedent/consequent, and cause/effect knowledge, such as "If the temperature exceeds 82 degrees Fahrenheit, then turn on the air conditioning." Frames are well-suited to represent descriptive and relational knowledge that clusters or that conforms somewhat to stereotypes, such as a description of a wildlife habitat or of an accounting process. Semantic nets are useful for modeling classifications, physical structures, or causal linkages, such as how various elements of an economic model influence each other. Of these knowledge representation forms, production rules are the most widely used for commercial applications at the moment.

An *inference engine* is a program's protocol for navigating through the rules and data in a knowledge representation in order to solve a problem. The inference engine embodies what we call deduction. The task of the inference engine is to select and then apply the most appropriate rule at each step as the expert system runs. This contrasts with conventional programming techniques in which the programmer selects the order in which the steps of the program execute at the time the program is written.

Some systems operate by chaining their inferences forward, some backward, and some utilize both directions. Backward-chaining systems begin with a goal, such as identifying a plant species, and work backward to seek a chain of premises that accounts for all of the facts at hand (such as botanical characteristics) that would arrive at a conclusion – the species name. Forward-chaining systems work from elementary pieces of information, such as a list of the components of a computer system, and attempt to build forward to a goal that will combine the elements, such as a printed configuration of components that represents a fully functional computer system. Most backward-chaining systems are used for diagnostic purposes; they begin with some goals (a list of symptoms to explain, for instance) and try to discover what premises would cause the symptoms to be true. Most forward-chaining systems are used for constructive purposes; they begin with a collection of premises and combine them to produce some end result. In most cases the choice of which direction the system should run in is made on the basis of how a person would solve the problem.

While it is possible to write production rule systems in any computer language with an "If...then....." construct, a specialized artificial intelligence language named OPS5, dedicated to production rules, is becoming increasingly popular. OPS5 implements both forward and backward chaining and has a built-in inference engine and pattern matcher. Most of the commercially successful expert systems have been written in the OPS5 language.

- *Expert Systems in the Real World*

To date, the largest and most intensively used commercial expert system is Digital Equipment Corporation's XCON (expert CONfigurer), formerly referred to as R1 at Carnegie-Mellon University where Professor John McDermott developed the initial 500-rule prototype. This system, written in OPS5, is used on a daily basis in Digital's worldwide manufacturing operations to configure most VAX and PDP-11 computer systems that are on order, prior to manufacturing them. Dennis O'Connor, who spearheaded the XCON development effort, reports that as of February 1985, XCON utilizes over 4,200 rules to accomplish the task, taking less than two minutes to configure most systems. In addition to making changes or corrections to ensure that necessary pieces of hardware and other components are included, XCON checks orders for correct cable lengths, power requirements, and many technical details and then prints a report with several diagrams to assist in assembly on the plant floor and at the customer's site. Engineers still configure orders that are not yet within XCON's range of knowledge, and knowledge continues to be added to XCON's knowledge base.

In addition to XCON and the successful expert system applications noted earlier in this chapter, there are systems that can, for example, help people write computer programs, assist in diagnosing a wide range of diseases, analyze investments, infer chemical structures from physical data, and assist in the analysis of structural engineering problems.

A problem lends itself to an expert system approach when

- A solution to the problem has so high a payoff that it warrants the development of a system: solutions are needed in the area, and other methods of obtaining them have not worked.
- The problem can be solved only by an expert's knowledge rather than by utilizing a particular algorithm, which traditional programming could handle.
- You have access to a willing expert who, with assistance, can formalize the knowledge needed to solve the problem. You need to interview an expert intensively in order to provide expertise to the system.
- The problem doesn't necessarily have a unique answer. Expert systems work best for problems that have a number of acceptable solutions.
- The problem changes rapidly (for instance, new components are continuously being introduced for computers that must be configured); or knowledge about a problem is constantly changing (as in the continuing discovery of causes and treatments of diseases); or solutions to problems are constantly changing (for example, new methods of equipment repair are being approved).

The development of an expert system is generally not considered to be finished once the system is brought online. Instead, expert systems tend to continue to be developed, with programmers updating the systems' knowledge and processing methods to reflect progress or modifications in the domain, or problem area, of the system. Because expert systems cannot yet learn on their own, they tend to require a continuing relationship with a human expert. Once the system can handle routine problems, the human expert can pursue more knowledge, which will eventually be shared with the system. Meanwhile, the system may make some novel inferences that enhance the expert's knowledge. A continuing partnership can thus improve the performance of both the person and the program.

Although expert systems can do a great many jobs, at this time they have limitations. With some exceptions, they cannot be used in situations in which realtime responses must be fast, say, for direct machine control. And expert systems of today are all special purpose, rather than general purpose, systems. They do not have humanlike "common sense"; that is, they cannot yet use knowledge of the world to generalize successfully. In addition, though some can tell you how they arrived at a conclusion, they can't evaluate the validity of that process.

Despite these limitations, it is important to note that an expert system need not solve all problems. Human experts can't solve all problems either, especially novel ones. It takes time to develop expertise, and that is a continuing process.

For more information on the implementation of expert systems, see Chapter 6.

Natural Language Systems

When we want to communicate with computers, we have to do it on their terms, in their language, and in their media. We have to learn Assembler, BASIC, COBOL, FORTRAN, or some other programming language. These languages tend to be unforgiving if we put a period in the wrong spot or leave out a parenthesis or some other minute detail. There are currently some attempts to alleviate the communication problem with interfaces such as menus, online help facilities, and graphic icons. These are still somewhat cumbersome, require some training, and tend to be very system- or application-dependent. Now, researchers are developing "natural language" systems that can accommodate our native tongues, such as English, Japanese, and French.

Research and implementation in natural language communication have addressed problems of input and output and how to get the computer to manipulate and respond to expressions in natural language. Natural language input and output are done through both text and sound. Communication via text is accomplished with traditional hardware mechanisms—keyboards, printers, and video monitors. Natural language communication via sound is accomplished with hardware mechanisms packaged as speech recognition and synthesis systems. Some of these recognition and synthesis systems are supported by AI technologies that interpret spoken input or generate humanlike audible speech output.

Conventions in discussing areas within the field of natural language are as follows. "Natural language" is an umbrella term for communication with computers using our native languages. It includes language input, output, and understanding. "Natural language" is also sometimes used in a more restrictive sense to refer to the text branch of the language problem. "Speech understanding" refers to the ability of computers to respond correctly to spoken language. And "speech generation" refers to the ability of computers to output spoken language.

As we will discuss later, some practical commercial systems are already available in two of the areas of natural language communication—speech generation and natural language interfaces. Other goals, particularly speech understanding, require a great deal more research before they will become widely used in commercial settings.

- *Language Understanding*

When we say that a computer "understands" language, we mean that it is able to process the plain language of the user, carry out the command, and generate appropriate output.

Uses for computers that can understand natural language include

- Interfaces to systems such as databases or operating systems, expert advisory systems, or robots.
- Machine translation systems to translate written materials from one language to another.
- Document-understanding systems that enable a computer to read and understand the information in documents well enough to summarize and redirect points of importance to various recipients and to organize and store information in order to answer questions on the contents.

Current Approaches to Natural Language Understanding

Two kinds of analyses are performed on input to natural language understanding systems. One focuses on syntax (how words are structured into expressions), and the other focuses on semantics (the meaning of words within the context of expressions). These analyses together allow natural language systems to generate a paraphrase of the input expression in an internal representation language.

Syntax—the role of each of the parts of a sentence—is analyzed by parsing. Parsing is the process of identifying what function each word in a sentence performs and whether the input sentence is "legal" (complete and grammatical), according to a particular grammar. A variety of grammars, supported by different theories about verbal communication, have been used to govern the parsing process in natural language systems.

Semantic analysis, which focuses on the meaning of words and phrases is then needed to clarify the system's understanding. Semantic analysis proceeds by associating words and their roles in an input sentence with information about the problem domain stored in the system's data or knowledge base. This knowledge can constitute a background context of expectations against which to interpret the input. The stored knowledge might be descriptions of objects, events, and relationships in the problem domain, descriptions of typical situations that might be encountered in the problem domain, or chains of events or procedures that occur given certain conditions in the problem domain.

In addition to using the information in the system's knowledge structures to analyze input, the system can store the input itself in knowledge structures, increasing and refining its knowledge as processing goes along. This helps the system to understand input in contexts larger than single sentences. The stored information provides a way to link references in one sentence to referents in another.

Natural language systems can combine information from the syntactic and semantic analyses to generate formalized representations of input sentences. The formalizations generated by the analyses can be stored in the knowledge base for comparison to other input or they may set off some activity in the system, such as answering a database query.

Obstacles to Machine Understanding of Language

Even with the support of a grammar, parsing is difficult without other kinds of information. People use their knowledge about the world to help clear up ambiguities in the parsing process. For instance, we human beings can extract sense from a sentence like "Sea otters crack mussels on rocks as they swim on their backs." We parse the phrase "as they swim on their backs" as modifying "otters" because we have knowledge about the world that tells us that rocks and mussels don't swim on their backs. (We would understand the sentence even though it is badly constructed.) A system without that knowledge would have trouble deciding how to parse the sentence. Other parsing ambiguities arise in words with double meanings, in idioms, and in pronouns. Semantic analysis helps to resolve these difficulties.

Ellipsis also presents difficulties to computers. Ellipsis occurs when words are left out, creating grammatical incompleteness. People interpret elliptical expressions from the context. As an example, if someone said, "A gang of five robbers held up the bank this morning," a reporter asking, "Any arrested?" would be using an elliptical expression that most listeners would understand.

Another problem for natural language systems is created by metaphors, like "Maria Garcia is a pillar of the community." Metaphors cannot be directly translated by a computer without explicit instruction.

Computer programs also have not yet been created that can read between the lines to interpret a message based on the goals of the communicator. For instance, if you are wrapping a package and ask someone, "Do you have any tape?" you are making a request, not taking an inventory.

Natural Language Interfaces

Natural language interfaces allow people to use subsets of their native languages to communicate with computers in restricted domains. It is expected that a major use of the technology will be in organizations that query databases. Managers, office workers, and technical professionals will be able to get needed data from the computer without going either to data processing courses or through the data processing department. This will make it worthwhile to pose ad hoc questions that may not warrant the development of full-fledged programs but that do contribute useful information to the problem at hand. A long-range goal is to make it easier and more comfortable for people to use their home computers and to program and control robots.

A few commercial natural language interfaces to databases are already on the market. One of the earliest of the commercial products was INTELLECT,† developed by Artificial Intelligence Corporation, Waltham, Massachusetts. INTELLECT is intended for use as a front-end interface to information retrieval applications in areas such as finance, marketing, manufacturing, and personnel. INTELLECT parses the user's natural language query into an internal representation that sets off a database search. To do this, in addition to using a grammar, the system also draws on knowledge of the database structure, database contents, a built-in data dictionary, and an application-specific dictionary. When a query results in the generation of more than one possible interpretive paraphrase, INTELLECT resolves the ambiguity by assigning preference ratings to the different paraphrases, choosing the one most highly rated due to its consistency with information in the database. If necessary, INTELLECT even asks the user which of several interpretations is correct.

An approach to interfaces that emphasizes semantics has been taken by Professor Roger C. Schank, chairman of Yale University's computer science department, director of Yale's Artificial Intelligence Laboratory, and founder of Cognitive Systems, Inc., New Haven, Connecticut. Cognitive Systems has developed natural language interfaces to databases and conversational advisory systems. Cognitive Systems' products "map" natural language input into conceptual representations based on Schank's theory of conceptual dependencies, which capture the meaning of the input. In addition to storing information about meaning in conceptual representations, these systems incorporate information about the problem domain in various knowledge structures, such as *scripts*. These knowledge structures aid interpretation by providing the system with expectations, contexts in which to understand input. A script, for instance, is a description of what happens in a situation that conforms to a stereotype. When a scripted topic is presented to a system employing this knowledge structure, it has a set of expectations that helps it resolve ambiguities.

Cognitive Systems' products also combine expert systems with natural language systems to form "intelligent retrieval systems." These natural language systems set up a context and keep track of it during a conversation. This adds to the system's fund of expectations, helping to resolve ambiguity. It is also possible to build into the system profiles of the users' goals, so that the system can retrieve not only the specific information requested, but also related information that would be of interest to the user. And again, the profiles provide a context that helps the system decide how to interpret a request: one person's "year," for instance, might be fiscal, while another's might be the calendar year.

Natural language interfaces vary in how much users must conform their input to the structure of the system, some allowing for the use of language that is more natural than others. Most allow users to modify the dictionaries with which the systems are equipped and to add their own entries, with varying degrees of ease and allowing varying kinds of definitions. Some systems handle sentences in which words have been omitted by proposing a fleshed-out interpretation to the user and asking if this was the correct reading. On encountering a spelling error, some natural language systems ask for a correction, some automatically correct the word, and others simply stop processing the sentence.

At this time, individual natural language interfaces must be specialized for particular subject matter contexts in order to interpret words and phrases correctly. HAM-ANS (Hamburg Application oriented Natural Language System), a German language interface developed by researchers at the Artificial Intelligence Research Institute at Hamburg University, West Germany, has been specially tailored for database access in a number of areas. One specialized use was to access a previously constructed database containing scientific information gathered on expeditions to the South Atlantic and Indian oceans. HAM-ANS was able to access the information without requiring modification of the database.

It should also be noted that today's natural language systems, and AI systems in general, are limited in knowledge, but the systems are usually not aware of these limits. Unlike people, the systems "assume" that they operate in a closed world, that their knowledge of the domain is complete and adequate. When a natural language system responds "no" to a query, it may really mean that the system doesn't know the answer. In reality, objects and relationships pertaining to most real-world domains change or are not modeled in the system. Another problem is that an appropriate response frequently depends on knowledge of the user's motivation in asking a question, and current systems are naive in this area.

Nevertheless, natural language interfaces are now available that successfully lessen communication obstacles to problem solving in some computing tasks and speed up access to information.

- *Machine Translation*

   Work in machine translation from one natural language to another has revealed that the subtleties of human language do not easily yield to computerization. The word-for-word translation systems of twenty years ago just didn't work. Research in machine translation has made it increasingly clear that human language cognition is a very complex ability that requires many kinds of knowledge, including knowledge of the structure of sentences; the meaning of words; the patterns of conversation; the expectations, goals, and beliefs of the partner with whom you're conversing; and a great deal of knowledge about the world as well as knowledge about the particular topic of conversation.

   Current implementations that most closely approach automatic translation may use syntactic and semantic information in order to translate words in context. Different systems require varying degrees of human assistance to edit machine-translated drafts or to assist in translating elements outside the bounds of the systems' abilities. In addition, systems described as "fully automatic" are, at this time, restricted to small domains. The speed of translation may be as slow as 600 words per hour for output that requires little editing or as fast as 60,000 words per hour for output that is likely to need considerable editing. Faster speeds are achieved in some systems by constraining the input to shorter sentences or by setting lower standards for the quality of the output. As with all natural-language systems, the more highly constrained the domain of discourse, the better the translation.

   The LOGOS† system, from Logos Corporation, Waltham, Massachusetts, is designed for business use. LOGOS works in partnership with a human translator. Before LOGOS begins a translation, the system examines the document for words it doesn't know. The translator then provides the system with information about those words, expanding the general dictionary. After the dictionary is complete for the purposes of the particular translation, LOGOS generates a draft of the document, which the translator then edits. Customizing the dictionary with multiple definitions of words for various contexts does not make the system unable to accept new releases of the vendor's dictionary.

Weidner Communications Corporation, Northbrook, Illinois, sells semiautomatic systems. The vendor supplies a core dictionary of 15,000 words and idioms to which users add their own terms. The Weidner system is much like LOGOS except that the machine prompts the translator interactively for words it doesn't know and incrementally builds a dictionary.

The ALPS† system from Automated Language Processing Systems, Inc., Provo, Utah, takes a slightly different tack in that the translation process operates in an interactive, rather than in a batch, mode. This system goes a bit further in that the interactive mode not only finds unknown words but also words with ambiguous meanings in context. In general, the ALPS dictionary is somewhat more sophisticated, accommodating word strings or phrases, such as idioms, as well as single words. Instead of one large dictionary, ALPS utilizes several reference dictionaries and, moreover, builds a separate dictionary for each document. The document dictionary can be fine-tuned for the document's specific context without affecting dictionary definitions that will be applied to documents written in other contexts.

Experimental systems are beginning to incorporate more semantic information in knowledge structures such as *conceptual dependencies* and *scripts* (see Chapter 3 for more information) that provide the system with knowledge that helps to resolve ambiguities in interpretation. For instance, semantic elements are being included in EUROTRA, a system being developed under the support of the European Economic Community.[1] This system, intended to perform machine translation among eight languages, is scheduled to be operational by the end of the decade.[2]

- *Document Understanding*

Researchers at Yale University, working under the direction of Professor Schank, have developed a number of document understanding systems. FRUMP, developed by Gerald DeJong (now associate professor of electrical and computer engineering, University of Illinois, Urbana), was an attempt to model how people skim newspaper stories. This system scans wire service stories and produces brief summaries of them in several languages. FRUMP utilizes "sketchy scripts," scripts that note only the most important aspects of a situation that conforms to a stereotype. A system called CYRUS, developed by Janet Kolodner (now associate professor of computer science, Georgia Institute of Technology) was designed to store information from FRUMP relating to the activities of former Secretary of State Cyrus Vance. The purpose of CYRUS was to model how people's memories are organized. The system cross-referenced information and was able to reorganize its memory structure to accommodate new information. CYRUS was aided in its task by the inclusion of a knowledge base about what secretaries of state do, about protocol, and other matters pertaining to Vance's activities.

CYRUS and FRUMP skim text, but BORIS is a story-understanding and question-answering system that involves many knowledge sources in an attempt to understand stories in as great a depth as possible. BORIS was developed by a team of researchers under the direction of Wendy G. Lehnert (now associate professor of computer science at the University of Massachusetts, Amherst) at Yale University. BORIS was an advance over modular systems. In BORIS, the various elements that process the input and aid understanding are integrated, not used one at a time. BORIS has four basic processing units that interact in order to understand a story in depth. A parser, called the conceptual analyzer, reads the English text and stores the information in conceptual dependency form. As the story is read in, the event assimilator examines the concepts stored up to that point in relationship to each other and to a previously stored fund of knowledge about the world. A question-and-answer module uses the conceptual analyzer to parse questions submitted to the system. The module then makes inferences based on memory contents, which the system expresses in conceptual dependency form. And finally, the English generator translates the conceptual dependency representation into English language output.

Document understanding systems such as those described above make it possible for computers to summarize text and generate responses based on content. They also make it possible for computers to store and retrieve information based on concepts, rather than just keywords.

- *Document Critiquing*

One aspect of document generation, text critiquing, has been implemented in EPISTLE.† EPISTLE was developed by researchers at the IBM† Thomas J. Watson Research Center. The current implementation of the system provides business letter writing support (spelling, grammar, and style checking) for office workers. A natural language processing unit analyzes typed text by means of an online dictionary and a system that parses sentences according to rules that encode English grammar. The system flags errors by highlighting the problem text and indicating the type of error (spelling, grammar, or style) in a *mode* window. The user selects an error to work on by pointing to it with the cursor. The system proposes corrections in a *fix* window. The user may implement a suggested correction, ignore it, request additional information, or substitute a correction of his or her own.

- *Speech Understanding*

Because we do not yet understand how human beings are able to make sense of the stream of sound that is spoken language, it is not surprising to find that this area of natural language communication is not yet in its maturity. Early techniques involved storing the sound patterns of a selection of words relevant to a problem domain and comparing the input signal with these patterns, attempting to make matches.

The Advanced Research Projects Agency (ARPA) of the U.S. Department of Defense sponsored the Speech Understanding Research (SUR) project in the early 1970s in an effort to promote advances in this field. HEARSAY-II, a document retrieval application developed by Carnegie-Mellon University in response to the ARPA challenge, was able to understand a 1,011-word vocabulary of connected speech from one male speaker after the system was supplied with about 60 training sentences pronounced by the speaker. HEARSAY-II understood the speaker's utterances with from 9 to 26 percent error.[3] HEARSAY-II is perhaps the best known of the SUR systems because of an innovative control structure that is not limited in applicability to speech-understanding: independent knowledge sources communicated with each other via a "blackboard" where results were shared and subproblems posed. This control structure had been used previously in HEARSAY-I, a speech-understanding system that played chess in response to voice commands.

Systems in use at the moment vary in several respects. Some are speaker-independent, while some recognize only a particular individual's speech. Some can recognize only isolated words, while others can pick a particular word out of a stream of connected speech, and some even understand connected speech within certain narrow limits. Systems also vary in the size of their vocabularies. Right now, there are speaker-dependent, isolated-word systems that can recognize about 1,000 words. For reliable recognition, that number falls to about 50 well-chosen words. A large vocabulary for a reliable connected-speech system would be 200 words, fewer still for a speaker-independent, connected-speech system.

To train a system to recognize words in a speaker-dependent format, you must provide it with samples of a person's speech. While there is some variation in the way one or more individuals pronounce consonants, there is huge variation in the pronunciation and diction speed for vowels. These factors require the training to include many varying samples.

Attempting to enable a system to understand connected, or continuous, speech adds difficulty to the problem. Syllables of adjacent words may blend or cause some sounds to be dropped. Since connected speech bears little resemblance to the stream of sound made if each word in the string is pronounced individually, it does not suffice to simply match patterns, word-for-word.

Difficulties for speech-understanding systems also arise with homonyms, words that sound the same and may or may not have the same spelling but have different meanings, such as

I heard the song.

I saw a herd of buffalo.

A closely related difficulty is presented by similar sounding phrases like "I scream" and "ice cream."

Connected speech is easier for a system to interpret when rules of conversation are provided to help predict which words can legally follow each other. Limiting the vocabulary to certain words within a domain also helps, reducing the processing time for pattern matching.

Research is now being done on providing systems with knowledge about the world that will help them predict what expressions might mean, based on context.

• *Speech Generation*

Speech generation is the term for a machine reading text aloud. The speech is the audible production of the output of a system, whose text has already been determined, in correctly pronounced speech. This part of the natural language problem has been solved with the arrival of commercially available speech-generation devices.

Digital Equipment Corporation has developed DECtalk, a product that converts alphanumeric text into human-quality speech. DECtalk uses logical rules to evaluate the context and punctuation of the phrase to be spoken and converts it to conversational English. It works by reviewing the whole input phrase to examine sentence structure, grammar, and context. DECtalk compares the incoming words to a dictionary of more than 5,000 exceptions, contractions, and abbreviations. If a match is found, the pronunciation is simply pulled from the list. If the word is not found in the main dictionary, DECtalk then searches through a second, smaller, user-defined dictionary that contains industry-specific words and abbreviations used in the particular subject area. Pronunciation for words not included in DECtalk's dictionaries is achieved by applying a set of 500 letter-to-sound rules.

A truly sophisticated speech generating system must pronounce phrases the way a human reader would. As an example, a simple device would read $125.75 as "dollars one two five point seven five." But state-of-the-art systems read it as it should be read: "one hundred twenty-five dollars and seventy-five cents." Conventional speech systems have pronunciations that are preset at the factory. But the incorporation of AI technologies allows for a flexible user interface in which the end user can specify details. An important feature is a choice of many natural-quality speaking voices with variable speaking rates and intonations, a choice between male and female voices, and other special effects.

✖ Robotics and Sensing Systems

The word "robot" does not mean the same thing to all people. In the United States, we limit the definition of robots to task-performing multifunctional manipulators that can be programmed and reprogrammed to do a variety of jobs. In Japan, the definition includes programmed manipulators that can perform only one task. Neither definition of robot necessarily includes artificial intelligence.

AI researchers have been working on ways to add intelligence to these machines in the form of vision, tactile sensing, planning, and learning. Even now, commercially available robots are capable of simple vision: Robovision® from Automatix, Inc., of Billerica, Massachusetts, for instance, is a vision-guided, arc-welding system. Simple tactile sensors are also available.

The refinement of advanced sensors and the incorporation of other AI technologies now being researched will eventually render robots capable of performing more sophisticated tasks. If robots had the ability to respond to natural language commands, they would be easier to control. Increased ability to process and respond to sensor and vision input would allow robots to manipulate objects more effectively and move more successfully through the environment. It is hoped that AI techniques will eventually enable robots to "learn," to solve problems based on changing needs in their task environments, and to plan how to accomplish tasks. These skills would make them useful in environments less structured than those in which they are used today.

Robots can be classified into three main groups—industrial robots, "android" robots, and automated environments (factories).

- *Industrial Robots*

The industrial robots are considered to be a type of machine tool, programmed to manipulate parts or tools in a sequence of motions. However, a term like "machine tools" does not fully describe or denote the growing sophistication of these products. More robot manufacturers are beginning to add vision, pressure, and tactile sensation capabilities to industrial robots.

At this time, most industrial robots have a single jointed arm with a gripper, and most of these robots are stationary. Because industrial robots are able to repeat programmed movements rapidly and precisely, they are good production-line machines. Industrial robots are used most prevalently in the metal-working sector of industry, particularly in automobile production. They are cost-efficient in hazardous or boring jobs like spray painting, welding, and the loading and unloading of machine tools. Industry utilizes robots for three important reasons. First, they can go where people can't go, into nuclear reactors, for instance. Second, because robots are precise and consistent, they turn out products of high and unvarying quality. And third, because robots, unlike people, can be kept free of dust and other particulate pollutants, they are ideal for working in environments such as the disk drive and semiconductor industries that must be totally free of contaminants.

Researchers are trying to give robots the capability of selecting single items from among many (an important ability for industrial bin-picking applications) and to manipulate parts intelligently in the environment in ways that are not preprogrammed. Researchers are also trying to increase the speed at which robots can respond to input from their sensors. The ability to brake quickly in response to obstacles, for instance, can be vitally important. Another area of research is "fixtureless assembly," a type of manufacturing in which robots hold and manipulate items with a high degree of freedom in contrast to the type of manufacturing that requires parts to be fastened down in precise positions.

• *Mobile Robots*

We have all read science fiction about robots who walk, but these android-style robots, as a class, are still in their infancy. Today's mobile robots do not look or act much like people. One robot has a central controller and radiating legs, a design that allows it to climb over many physical obstacles. A hopping machine with just one springy leg has also been built and demonstrated at Carnegie-Mellon University by Marc Raibert (now associate professor of robotics, Carnegie-Mellon University) and his colleagues. And yet another type of mobile robot, useful for transporting items, looks more like a small boxcar than a person. Examples of the latter type of robot are automatically guided vehicles (AGVs) that bring mail to offices or parts to factory workstations. The more sophisticated AGVs are programmable, can onload and offload, and can be controlled by radio.

Plans are under way to provide mobile robots with refined wheels and jointed legs. In order to navigate around some obstacles, robots need more mobility than mere wheels can offer. Odetics, Inc., of Anaheim, California, has developed a six-legged robot called a functionoid. This robot is stable on three legs, can use its legs as arms, and can climb steps 33 inches high and maneuver through doorways.

Navigation is no problem when the robot's route is unchanging and free of obstacles. Slow travel on wheels is now guided by wires or chemical paths set out on or in the floor. But when it comes to navigating intelligently through unfamiliar terrain or among obstacles, robots must rely on sensing systems (sonar, radar, tactile, and capacitance). It is anticipated that improved vision and sensing systems will enable robots to move more quickly and safely through complicated environments.

- *Flexible Manufacturing Work Cells*

Flexible robotic manufacturing environments are another area of research in which AI methods are being used. The goal is to create systems that can be quickly and inexpensively modified to produce variations in products as well as totally new products. This will be useful in high-mix, medium-volume batch operations. In Japan, the government is sponsoring a team to build a completely automated factory that will require only two or three people to run it. The United States National Bureau of Standards also has a prototype of such a factory at its Automated Manufacturing Research Facility in Gaithersburg, Maryland. The United States Air Force's Integrated Computer-Aided Manufacturing (ICAM) project plans to build a sheet-metal fabrication center scheduled for completion in 1985.

- *AI and Robot Teaching Methods*

Artificial intelligence-based methods for teaching robots are currently under research. One method involves entering the endpoints of a motion into a computer that automatically programs the full motion into the robot's memory. This offline programming allows for more refined, complex motions than are possible with the methods commonly used today. It is expected that one day speech recognition systems will be incorporated into robots that will allow their users to give vocal commands to them. And eventually, it is anticipated that AI will enable robots to learn from experience and work out how to perform unfamiliar tasks on their own. Some progress was made in this area with the STRIPS program (Stanford Research Institute Problem Solver), written by Drs. Richard Fikes (now with Intellicorp) and Nils Nilsson (now chairman of the computer science department, Stanford University). Given a map of the environment and an internal representation of the goal, STRIPS allowed a robot in a simplified environment to work out a plan to change its location, move an object, and store the plan for future use.[4]

- *Vision Systems*

Vision systems perceive the location, form, size, shape, or color of objects. They process and interpret images in order to analyze and identify them. Vision systems have already proven useful in natural resources analysis for remote sensing; in manufacturing for parts sorting, assembly, and quality control; and in medicine for internal examination of the body. Sorting, assembly, and inspection systems are used in conjunction with robotics today, but research is aimed at providing more advanced vision systems that will further expand the capabilities of robots by improving their ability to navigate through complicated environments and to manipulate objects.

Vision systems, in general, were originally considered to be part of AI research. However, because "low-level" vision systems are not generally "knowledge-based" in the current sense of the term (equipped with knowledge stored in rules so it can be used flexibly in unanticipated ways), they are not considered, in today's view, to employ AI methods. These low-level systems measure various features of images and, in some cases, can be efficiently implemented as hardware.

"High-level" vision systems, which do employ AI symbolic reasoning techniques, must cope with ambiguities. For instance, they can identify objects as instances of models by comparing aspects of the objects to stored parameters, and they deal with problems of occlusion (where an object's shape appears to be different from a model because part of it is hidden). The high-level systems are equipped with heuristics (knowledge or rules of thumb) about the size, shape, and spatial relationships of features, to interpret images. Expectations about features in the object or scene and knowledge of the purpose of the visual processing also help high-level systems interpret images. High-level systems with AI capabilities have not yet been implemented in the high-speed manner provided by hardware. Instead, high-level systems are generally implemented as software, which allows for the currently needed flexibility.

There are two- and three-dimensional computer vision systems. Two-dimensional computer vision systems are those that interpret two-dimensional objects or scenes, such as aerial photographs or microscope cross-section images. Three-dimensional computer vision systems, which may rely on heuristics and constraints to interpret the depth dimension, are required for three-dimensional objects or scenes, such as bin-picking applications (a common industrial task) or accurate assembly line work.

Devices that sense two-dimensional information can be used for input to both two- and three-dimensional vision systems. A sensor that directly reads the depth dimension (an activity referred to as "active sensing") is not always necessary to look at the world that is interpreted by a three-dimensional system. A single camera, for instance, can record a three-dimensional scene; its depth can then be estimated by means of knowledge about angles, shadows, and perspective. Active sensors that directly and accurately measure depth include laser range finders, structured light systems, and sonar systems. Laser and sonar range-finding devices measure distance by timing the reflection of light or sound from the source to the object and then back again. Structured light systems (in which light is emitted in a particular pattern, such as "Venetian blind" stripes) interpret the patterns formed by a light beam's interference with an object's surface by triangulation. Laser scanners are used for analyzing images on film, sonar devices have been used for mapping the topography of the sea bed, and structured light systems have been used for inspecting objects on assembly lines.

Research is now in progress on binocular three-dimensional vision, like that of human beings and many other animals. One approach to modeling human binocular vision pursued at the Massachusetts Institute of Technology involves computing the parallax between images received by two camera "eyes." The views of the two cameras are necessarily different, and objects seen in one view may even be occluded in the other. The objects seen by the two cameras must be compared and matched with each other before the parallax can be measured. Artificial intelligence techniques are used for this purpose.

Systems are available that code vision input as merely dark-and-light values, as a fuller range of shades of gray, or even as various colors. Dark-and-light or "binary" systems are adequate for tasks that require only the recognition of outlines or basic shapes. Gray-scale coding is necessary for tasks that must recognize features within an outline or shape.

For a sensing system to interpret a scene or object, it must first receive input, which frequently comes from a videocamera. Alternative input devices include the CCD (charge-couple device) and laser scanners. A digitizer converts the image from analog signals in the form of voltages into digital code, which can then be processed by computer programs. The next processing step—feature detection—applies computer pro-

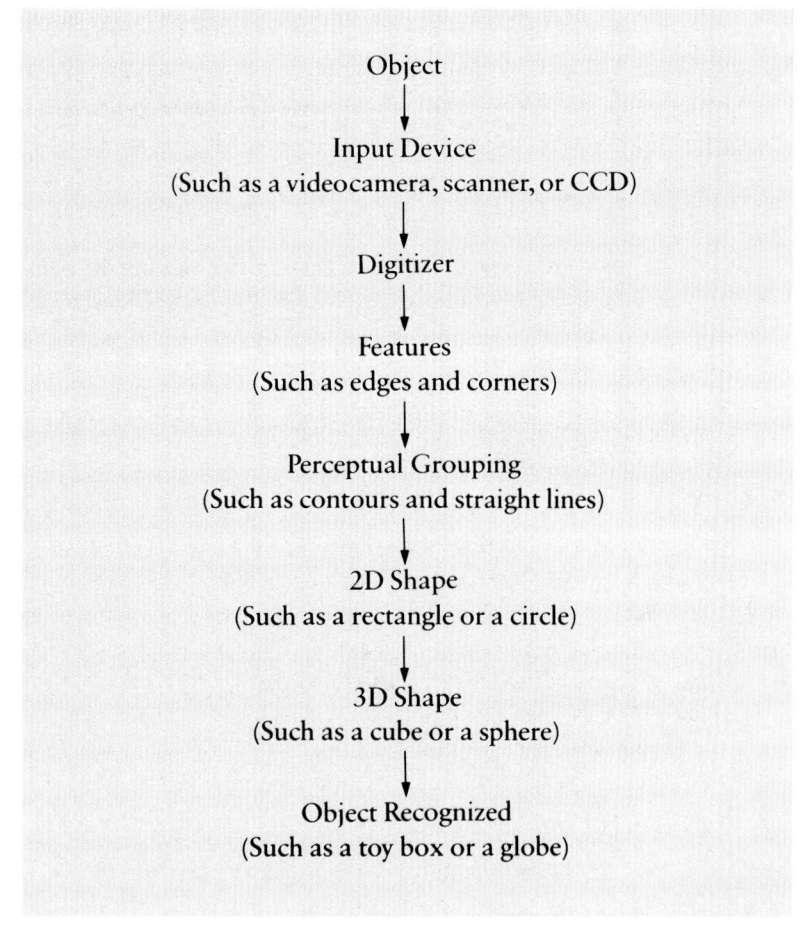

Object
↓
Input Device
(Such as a videocamera, scanner, or CCD)
↓
Digitizer
↓
Features
(Such as edges and corners)
↓
Perceptual Grouping
(Such as contours and straight lines)
↓
2D Shape
(Such as a rectangle or a circle)
↓
3D Shape
(Such as a cube or a sphere)
↓
Object Recognized
(Such as a toy box or a globe)

These are the basic processes involved in vision systems.

grams to the digitized signal to extract edges, corners, and regions. The perceptual grouping process applies other computer programs to identify features that belong together in particular visual structures.

Qualities useful in identifying coherent regions of the image are texture, shading, distance, orientation, and reflectance. When these qualities of the image are abstracted and stored with the original image, they are called intrinsic images; as simplifications, they are more usable in the pattern-matching that identifies an object. Statistical methods may be used to determine how closely the image under analysis compares to parameterized models of objects. If no model of an object has been stored, the system will be unable to recognize it.

In a 2D system, this pattern-matching is the final step. In a 3D system, one more step is necessary; computer programs must apply heuristics and knowledge about the physical world to interpret the depth information about the image from clues provided by variations in shading, texture, or surface contours.

Most vision systems include methods for teaching the computer about new objects. This may be done by showing the object to the camera in a number of orientations or by storing statistics about the form of the object.

- *Touch Sensing*

Touch sensing evaluates objects on the basis of weight, shape, texture, vibration, direction, pressure, and temperature. Robots are expected to benefit from touch sensing because it will give them information about areas they cannot see when their grippers are in the way. Touch will also allow for the more precise control needed in grasping delicate objects. It is anticipated that proficient tactile sensing will also decrease the current need for expensive fixturing, which must today keep robots and parts precisely in position for tasks. Tactile sensing will allow a robot to accommodate to alterations in its own position or the positions of parts, and to respond to sensed factors in its environment, such as temperature, by making appropriate adjustments.

Some robotic sensing systems are simply hardware mechanisms without AI content, such as limit-switch mechanisms or force/torque measuring devices. Some tactile sensors employ an array of transducers, each of which converts force or displacement information into electrical signals. The array may be coded as "binary" (each bit is either on or off) or "pressure sensitive" (each transducer has a value corresponding to the amount of pressure exerted on it). This array of signals represents an image of the object contacted, analogous to either a binary or a gray-scale image from a vision system. The array provides an interpretation of shape and pressure that can be used to calculate how to grasp an object properly. One medium for detecting and then relaying sense information is a skinlike sensing surface that generates electric currents when deformed. A number of artificial skins under development can receive and transmit contact impressions, including those of pressure and temperature. Potential materials for such skins include a ceramic conductive rubber, PZT, and polyvinylidene fluoride (PVF_2),[5] a plastic.

Touch sensing, like vision, is a kind of image processing. And like vision, it makes use of image enhancement, analysis, and identification. As in vision systems, there is a low-level and a high-level approach. Using the low-level approach, the sensor receives an impression referred to as the image, features are extracted from it, and then a model is developed that interprets the object image. In the high-level approach, the system begins with a hypothesis of what the object is, and this hypothesis dictates which experiments—such as moving the sensor to obtain certain measurements—the system should run in order to verify the hypothesis. The high-level approach involves AI techniques. If tactile sensing can one day be combined with vision sensing and the problem-solving aspects of intelligence, robots will become much more versatile. The coordination of the various AI technologies will require powerful control capabilities, supported by effective programming languages.

✳ Automatic Programming and Intelligent Programming Aids

Automatic programming is the application of artificial intelligence techniques to the general goals of automatic program construction and automatic program transformation. In order to achieve these goals, in most cases it is necessary for automatic programming systems to incorporate two kinds of knowledge—knowledge about general programming principles, including the programming language that will be used and the hardware and operating system on which the program will run; and knowledge about a specific application domain.

Knowledge about the problem domain is as necessary to produce an efficient program as is knowledge about programming. Domain knowledge enables the system to make use of constraints on the resulting program imposed by the application. Domain knowledge also helps a system choose the most efficient implementation method for a particular problem.

The area of automatic programming comprises a number of different disciplines, such as automatic program synthesis, automatic program verification, and optimizing compilation. The object of automatic program synthesis is to construct programs from rigorous and nonalgorithmic specifications describing what the programs should do. The purpose of automatic program verification is to prove the correctness of programs. This goal is achieved by using mathematical techniques to show that programs behave according to their formal specifications. Optimizing compilers attempt to make the machine-executable code more efficient by using search techniques and building knowledge bases. The compiler applies search techniques to find patterns in the use of variables, and the knowledge base is used to store especially useful segments of code.[6]

- *Automatic Programming and the Programmer*

One aim of automatic programming is to eliminate the errors and loss of information that occur in creating, documenting, and maintaining software by incorporating intelligent programming systems and aids into computer users' systems. Program synthesis systems, for instance, should greatly improve programmers' productivity and the reliability of their products. These systems will also allow programmers to work more on the initial system design and less on maintenance.

Automatic programming may also make it possible for nonprogrammers to use computers to solve problems. Automatic code generation is one aspect of the task. Another is the development of intelligent programming aids, such as an automated help facility that will respond to a user's natural language query with an answer or by carrying out the user's request.

An advantage of automatic program synthesis is that the programmer can express the purpose of the program in specifications that are close to the way he or she thinks without having to make them computable.

Another advantage is that maintenance need not be done on the source program but at a higher level — on the specification. This simplifies the user's maintenance task. The user must first make a few choices about how the program should be implemented, but then can leave the coding up to the automatic programming system. In addition to freedom from clerical errors, increased opportunity for optimization, and better documentation, automatic programming makes possible a library of reusable software, a library of specifications, rather than implementations that have a low potential for reuse. Such a programming assistance system, capable of carrying out the lower-level development tasks, will require a great deal more research in the formalization of development activities and in developer/assistant interfaces.[7]

- *Approaches to Automatic Programming*
Approaches to automatic program synthesis have included

- Theorem-Proving Based Synthesis. This approach involves building programs in terms of mathematical theorems and verifying each step by proof. This method guarantees the correctness of the program, but it has been successfully applied only to a rather limited set of problems.
- Deduction-Driven Synthesis Based on Transformation Rules. The system develops a program from the user's specifications by means of fully defined transformation rules. The correctness of programs generated by means of these rules can be guaranteed.
- Knowledge-Based Synthesis. The system translates a user's informal queries into formal statements, which it can then convert into algorithms. Then the system selects the programming techniques, encoded in the knowledge base, that most efficiently accomplish the specification of the algorithms and maps them into a total description of the program that is desired. And finally, the system codes the algorithms in the target computer language. This approach, which has often been combined with natural language interfaces, is the one most identified with the name "automatic programming." However, the correctness of programs generated by the ad hoc rules of this method cannot always be guaranteed by the automatic program synthesizer.
- Problem-Reduction Based Synthesis. In a process called task decomposition, the synthesizer decomposes the original problem into sub-problem specifications, which are assumed to be simpler, until elementary specifications are obtained.

- *Applications*

It should be noted that on a certain level the LISP language has an inherent capability to express the kind of computations required by automatic programming. A piece of data can be viewed as a piece of program and, conversely, programs may be viewed as data structures in the language. This makes it possible for LISP to read, manipulate, and write programs written in the LISP language. Someone working in the LISP environment can run programs generated in it without first having to exit.

USE.IT   Higher Order Software, Inc., Cambridge, Massachusetts, has utilized the theorem-proving based approach in a software engineering tool called USE.IT.† USE.IT, which runs on Digital Equipment Corporation VAX superminicomputers and on IBM mainframes, creates correct FORTRAN and Pascal programs. It helps the user generate a logically sound flowchart of the proposed system.

As the user builds a system design, USE.IT applies mathematical theorems that determine whether an error has been made. As soon as an error is detected, an error message appears on the screen. In addition to automatically generated code, USE.IT provides automated documentation—a summary of the flowchart and complete listings of the source code and the hierarchy of modules.

The software engineer using this tool performs more design work and less coding drudgery than is usual.

CHI   CHI, developed at Kestrel Institute, Palo Alto, California, under the direction of Dr. Cordell Green, the institute's director, is an example of a deduction-driven system based on transformation rules. It is a prototype of a highly integrated intelligent programming environment. On completion, CHI is intended to produce error-free code. CHI consists mostly of programming knowledge expressed as production rules written in V, a very high-level programming language.

The V language integrates several programming paradigms—mathematical concepts such as sets, sequences, and relations; first-order predicate calculus; Ada®- and Pascal-like control and data structures such as loops, procedures, arrays, and vectors; and production rules.

CHI selects the appropriate transformation rules to implement the user's specifications, producing lower-level code, like LISP. The CHI system even compiles and recompiles itself, growing constantly "smarter" in a self-applied programming cycle.

Reasoning Systems, Inc., Palo Alto, California, is developing a commercial knowledge-based set of software tools embodying many of the key ideas from the CHI research prototype.

GIST Another deductive transformation-rule system under development utilizes GIST as a program specification language. Developed at University of Southern California Information Sciences Institute, Los Angeles, California, GIST grew out of automatic programming work of the late 1970s in formalizing informal programming processes. This work was rooted in natural language research. Researchers found they could provide logically formal constructs that appeared close to the way people think. These constructs were not suitable for compilation on traditional computer architectures, but they could be compiled using tools written in the AI language of LISP. GIST's library of rules transforms a fairly natural human way of stating a problem into more executable statements.

The Programmer's Apprentice The Programmer's Apprentice (PA), under development at the Massachusetts Institute of Technology, is a software development system that relies on what its designers refer to as "programming clichés." Believing that we are a long way from achieving completely automatic programming systems, Charles Rich and Richard C. Waters, principal research scientists with MIT's Artificial Intelligence Laboratory, designed the system as an intermediate solution—"an expert programmer's junior partner and critic." On completion, the PA will allow a person to build a program from clichés, fragments corresponding to common algorithms, and represented in the system by "plans." These plans represent only the essential features of algorithms. Rich and Waters believe that several thousand plans will be required to cover the domain of fundamental programming clichés to a useful extent. In addition to plans, the PA will be equipped with some knowledge of the problem domain.

The current demonstration system is a knowledge-based editor that uses a small number of plans to build up a program or to modify one in terms of its logical structure. In using the system, the programmer's job is to maintain an overall view of what must be done to accomplish the goals of the program. The editor's job is to keep track of the details of the implementation of the existing program. An advantage of using this system is that the opportunity to make many simple errors is eliminated for both the system and the programmer because the programmer works not on code but on the plans, and only the coding module of the system must work on the level of code. The programmer can also construct programs faster by drawing from the stock of plans in the system's library.

BIS The BIS (BIdirectional synthesizer) System, developed by Marco Somalvico (professor of computer science), Ugo Buy (now a software engineer at Digital Equipment Corporation) et al. at the Artificial Intelligence Project, Milan Polytechnic University, utilizes the problem-reduction approach. This technique, used in a variety of contexts in AI, consists of decomposing a complex problem into simpler subproblems. The decomposition process is repeated until it produces elementary subproblems having solutions that are known to the system.

In BIS, the task of program construction is viewed as problem-solving activity consisting of a top-down and a bottom-up phase. A correspondence is established between a problem to be solved and a program to be synthesized. During the top-down phase, program specifications are decomposed until elementary specifications are reached whose solving pieces of code are known. In the bottom-up phase, the elementary program segments are assembled to build up the final program.

In this chapter, we have seen the areas in which most of the current practical applications are being developed. In the next two chapters, we will consider more closely the techniques and languages used to implement them.

❊ Annotated Bibliography

Albus, James S., "Industrial Robot Technology and Productivity Improvement," Industrial Systems Division, National Bureau of Standards, Appendix B in *Exploratory Workshop on the Social Impacts of Robotics Summary and Issues, A Background Paper*, United States Office of Technology Assessment, Washington, D.C., February 1982. This paper discusses technical problem areas, the history of the field, current research and commercial applications, and future expectations. It provides an optimistic view of social impacts.

Ayres, Robert, and Steve Miller, *The Impacts of Industrial Robots,* Carnegie-Mellon University, Pittsburgh, 1981. This report briefly describes robot technology and goes into more depth about uses of robots and some of the anticipated social and economical impacts of their use. Human resource policy issues are discussed. A chronology of robotics technology is also given.

Barr, Avron, and Edward A. Feigenbaum (eds.) (Vol. I and II) and Cohen, Paul R., and Edward A. Feigenbaum (Vol. III) (eds.), *The Handbook of Artificial Intelligence,* 3 Volumes, Stanford, California, HeurisTech Press, Vol. I, 1981, Vols. II and III, 1982, copyright by William Kaufman, Inc., Los Altos, California, 1981. Technical aspects of expert systems, natural language systems, robotics and vision, and automatic programming are among the topics discussed.

Barstow, David, "A Perspective on Automatic Programming," *The* AI *Magazine,* Spring 1984. Barstow discusses the role of knowledge of the task domain, illustrated in an oil well log interpretation application.

Boden, Margaret A., *Artificial Intelligence and Natural Man,* New York, Basic Books, Inc., 1977. Boden's book describes artificial intelligence in a way that stresses its human relevance, particularly in relation to psychology and philosophy. Boden uses plain language and makes the workings of natural language programs clear.

Buy, Ugo, Francesco Caio, Giovanni Guida, and Marco Somalvico, "BIS: A Problem-Solving Based Methodology for Program Synthesis," *Conference Proceedings of the Second Annual Meeting on Artificial Intelligence,* Leningrad, USSR, November 1980. This paper provides a short historical survey of program synthesis systems and a detailed explanation of the Bidirectional Synthesizer System. It concludes with a discussion of future research directions.

Davis, Randall, and Douglas B. Lenat, *Knowledge-Based Systems in Artificial Intelligence,* New York, McGraw-Hill, Inc., 1982. This book presents issues of knowledge representation, utilization, and acquisition in terms of the mathematical expert system AM and the knowledge engineering system TEIRESIAS, which applies metalevel knowledge.

Edson, Dan, "Vision Systems for Bin-Picking Robots Increase Manufacturing Options," *Mini-Micro Systems,* July 1984. This article tells how vision capabilities are being integrated into robots to help them perform complex tasks.

Feigenbaum, Edward A., and Pamela McCorduck, *The Fifth Generation,* Reading, Massachusetts, Addison-Wesley Publishing Co., 1983. In addition to discussing the race to build a fifth-generation computer, this book discusses uses for expert systems.

Ferrai, D., M. Bolognani, and J. Goguen (eds.), *Theory and Practice of Software Technology,* North-Holland Publishing Company, 1983. The article "Operational Specification as the Basis for Specification Validation," which discusses GIST, is included.

Fu, K.S., and Azriel Rosenfeld, "Pattern Recognition and Computer Vision," *IEEE Computer*, October 1984, pp. 274-282. This article provides a brief description of the state of the art in computer vision.

Heer, Ewald, "Robots and Manipulators," *Mechanical Engineering*, November 1981, pp. 42-49. This article provides an overview of industrial robots — what they are, how they are used, how they are taught, and future uses for them.

Heidorn, G.E., K. Jensen, L.A. Miller, R.J. Byrd, and M.S. Chodorow, "The EPISTLE Text-Critiquing System," *IBM Systems Journal*, Vol. 21, No. 3, 1982. This paper discusses the initial objectives of the EPISTLE system and describes the user interface, some details of the implementation, and the three levels of processing.

Hendrix, Gary G. (ed., contributor), *Tutorial No. 3 on Natural Language Processing*, 1983 AAAI Convention held in Washington, D.C., (Copyright held by Gary G. Hendrix.) This document contains articles on various aspects of natural language.

Hillis, William Daniel, "Active Touch Sensing," MIT Industrial Liaison Report, AI Memo 629, April 1981. This report describes the construction and capabilities of a tactile sensor array and a mechanical finger, describes a program that uses the finger and sensor to recognize small objects, and provides an appendix on tendon hands and arms.

Jan Johnson, "Easy Does It," *Datamation*, June 15, 1984, pp. 48-60. This article provides a comparison of commercial natural language query systems.

Keller, Erik L., "Clever Robots Are Set to Enter Industry En Masse," *Electronics*, November 17, 1983, pp. 116-129. This article discusses the robotics market and vision and sensing systems. It provides a comparison of United States and Japanese robotics.

Kinnucan, Paul, "How Smart Robots Are Becoming Smarter," *High Technology*, September-October 1981, pp. 32-40. This article discusses vision and sensing systems and manipulator design.

Korf, Richard E., *Space Robotics*, Carnegie-Mellon University, Pittsburgh, 1982. This report surveys the possible applications and technical feasibility of robots in space. The future of the space program in the timeframe of 1980-2000 and critical technologies needed to support the space program are assessed. A research program is outlined for the development of autonomous space robots.

Lehnert, Wendy, Michael G. Dyer et al., "BORIS — An Experiment in In-Depth Understanding of Narratives," Yale University Department of Computer Science Research Report #188, January 1981. This is a report on a story-understanding and question-answering system that involves the specification and interaction of many sources of knowledge.

Lundquist, Eric, "Robotic Vision Systems Eye Factory Applications," *Minimicro Systems*, November 1982, pp. 201-210. This article discusses vision technologies, vendors, and products.

National Research Council, "State of the Art for Robots and Artificial Intelligence: A Summary," *National Productivity Review,* Autumn 1984, pp. 375-381. This is a reprint of the appendix to a 1983 study titled *Applications of Robotics and Artificial Intelligence to Reduce Risk and Improve Effectiveness* by the Committee on Army Robotics and Army Intelligence. In this reprint, two of the table heads have been modified to update the timeframes.

Parker, Richard, and Nicolas Mokhoff, "An Expert for Every Office," *Computer Design,* Fall 1983, pp. 37-46. This article describes how knowledge-based systems can improve office productivity by applying artificial intelligence to routine office tasks.

Raibert, Marc H., H. Benjamin Brown, Jr., and Michael Chepponis, "Experiments in Balance with a 3D One-Legged Hopping Machine," *The International Journal of Robotics Research,* Vol. 3, No. 2, summer 1984, pp. 75-89. This paper describes experiments that demonstrated that a 3D one-legged hopping machine was surprisingly simple to implement and powerful enough to permit hopping in place, running, and travel along a path.

Rauch-Hindin, Wendy, "Natural Language: An Easy Way to Talk to Computers," *Systems & Software,* January 1984, pp. 187-230. This article explains concepts on which natural language systems are based and describes some commercially available systems.

Research Resources Information Center, *The Seeds of Artificial Intelligence*: SUMEX-AI, Bethesda, Maryland, Division of Research Resources, National Institutes of Health, U.S. Department of Health, Education, and Welfare, Public Health Service, NIH Publication No. 80-2071, March 1980. This report provides a history of AI within the larger context of the history of computing and a description of the purposes and projects of Stanford University Medical School's SUMEX-AIM. SUMEX-AIM is the nucleus for a national community of biomedical AI projects.

Rich, Charles, and Richard C. Waters, "Abstraction, Inspection and Debugging in Programming," A.I. Memo No. 634, June 1981, Massachusetts Institute of Technology Artificial Intelligence Laboratory. This paper describes goals remaining in a 5-year research project to study the principles underlying the design and construction of large software systems and to demonstrate the feasibility of the Programmer's Apprentice.

Rich, Elaine, *Artificial Intelligence,* New York, McGraw-Hill Book Company, 1983. This is a highly readable introductory textbook on the techniques and implementation areas of AI. Rich provides clear explanations of how knowledge representations are implemented in the computer.

Rich, Elaine, "The Gradual Expansion of Artificial Intellligence," *Computer,* May 1984, pp. 4-12. Rich's article discusses automatic programming and automated HELP facilities.

Schank, Roger, Janet Kolodner, and Gerald DeJong, "Conceptual Information Retrieval," Yale University Department of Computer Science Research Report #190, December 1980. This report cites the CYRUS and FRUMP systems in discussing the need for natural language understanding capabilities in intelligent information retrieval systems.

Thames, Cindy, "Electronics' Newest Labor Force: The Steel-Collar Worker," *Electronic Business*, June 15, 1984, pp. 102-103. This article discusses the use of robots in the electronics industry, including vendors and prices.

Tucker, Allen B., Jr., "A Perspective on Machine Translation: Theory and Practice," *Communications of the ACM*, April 1984, Vol. 27, No. 4, pp. 322-329. This article discusses three "generations" of machine translation systems, citing examples and discussing future directions.

Waltz, David, L., "Artificial Intelligence," *Scientific American*, October 1982, pp. 118-133. This is a good introductory article to the field.

Waters, Richard C., "The Programmer's Apprentice: Knowledge-Based Program Editing," *IEEE Transactions on Software Engineering*, Vol. SE-8, No. 1, January 1982, pp. 1-12. This paper describes an intitial implementation of an interactive programming assistant system called the Programmer's Apprentice.

Winston, Patrick H., *Artificial Intelligence*, (2nd ed.), Reading, Massachusetts, Addison-Wesley Publishing Company, 1984. Winston discusses basic theoretical concepts of AI, citing successful applications. Among other AI problem-solving techniques, the author discusses natural language, vision systems, and robotics.

Winston, Patrick H., and Karen A. Prendergast (eds.), *The AI Business*, Cambridge, Massachusetts, The MIT Press, 1984. This book shows the reader how AI techniques have solved some practical problems and presents discussions of business issues involved in AI technology.

Notes

1. Tucker, Allen B., Jr., "A Perspective on Machine Translation: Theory and Practice," *Communications of the ACM*, April 1984, Vol. 27, No. 4, p. 323.

2. Ibid.

3. Barr, Avron, and Edward A. Feigenbaum (eds.) (Vol. I), *The Handbook of Artificial Intelligence*, Stanford, California, HeurisTech Press, 1981, pp. 327-331.

4. Rheingold, Howard, "The Well-Tempered Robot," *Psychology Today*, December, 1983, p. 44.

5. Keller, Erik, L., "Clever Robots Set to Enter Industry En Masse," *Electronics*, November 17, 1983, p. 120.

6. Rich, Elaine, "The Gradual Expansion of Artificial Intelligence," *Computer*, May 1984, p. 5.

7. Balzer, Robert, Thomas Cheatham, Jr., and Cordell Green, "Software Technology in the 1990s: Using a New Paradigm," *IEEE Computer*, November 1983, pp. 39-45.

Chapter 3

**The Techniques
of Artificial Intelligence**

No strict division exists between techniques used in artificial intelligence and those used in conventional computing. Some problem-solving techniques such as timesharing, interactive editing and debugging, and search that originated in AI research labs have entered the mainstream of computer science. By the same token, many artificial intelligence programs use conventional algorithmic approaches for specific tasks. However, the field of AI does offer some distinctive approaches to complex problem solving. This chapter provides an overview of the techniques that characterize problem solving in artificial intelligence. It also discusses how they differ from the methods that predominate in conventional computing.

AI Problems Call for Unconventional Problem-solving Methods

Symbolic Processing versus Conventional Computing

In conventional computing, a programmer creates computer program instructions that follow one solution path for each situation the program is designed to handle. The solution path is completely planned by the programmer—any surprises in the events of processing are bugs that must be eliminated. This conventional method of using predictable steps can be very powerful. It enables us to solve problems that require the processing of large amounts of data and the repetition of many steps. Over the years, progress has been made in conventional computing chiefly in two areas. First, programs can be executed faster today with improved hardware. Second, the time and effort required for writing the programs have been reduced by the development of more sophisticated instruction sets and programming languages that require the programmer to write fewer lines of code.

However, many important problems are not readily solved by conventional computing, no matter how fast it is performed. These problems involve decisions based on the complex interaction of many factors that must be considered as a whole, rather than as a series of steps. But artificial intelligence uses a technique that is helpful in solving these complex problems—symbolic processing. The symbols processed by artificial intelligence programs often represent real-world entities and, instead of simply performing calculations, AI programs manipulate, or "think about," the relationships among the symbols.

We could say that both conventional computing and artificial intelligence computing involve symbol processing. In traditional data processing, the system processes the content of variables. AI systems can do this, but they can also manipulate symbols independently of their values. This makes it possible to solve a problem when the value of a variable is not known until shortly before the answer is needed. For instance, an automobile that is being built must have a color, but the value of that color isn't important until delivery. It's difficult to write conventional software that allows variables with no value (remembering that 0 is a value), but symbolic processing allows a symbol to represent the car's deferred color assignment. An artificial intelligence system can manipulate such symbols independently of their values.

In data processing, it is the programmer, not the machine, who determines all of the relationships among the symbols. But in AI symbol processing, the program can determine relationships among the symbols that were not made explicit by the programmer. An artificial intelligence program can do this because it has rules for manipulating relationships among symbols whose meanings have been represented within the program by the programmer. This manipulation of relationships among pieces of data is an important feature of artificial intelligence programming.

Similarly, while you can store relationships in a conventional database, you can output only the types of relationships that you put in. The conventional program can manipulate those data with various algorithms, such as the formulas and calls of FORTRAN. However, the FORTRAN program will not establish new relationships—as relationships—among the pieces of data in the database. It will simply perform operations on the data in order to come up with new values. Even a relational database, which can reorganize data, must have its input categories ("attributes and rows") and its output organization specified by the programmer.

- *AI Programs Store Knowledge, Not Just Data*

To shed more light on the importance of the storage of relationships, let's make some distinctions among data, information, and knowledge. We can think of data as any value available to the system for processing. Information can be described as data that has been selected and organized for a particular purpose. Knowledge, in the realm of artificial intelligence, is information structured in a way that brings out and exploits the relationships among the pieces of data.

The distinctive aspect of the AI approach is the emphasis on the storage and manipulation of the relationships among symbols. So, artificial intelligence technology is concerned with preserving not just data but also the knowledge embodied in the relationships among the data. Statements expressing particular chunks of knowledge are then treated as data in AI programs.

- *AI Makes Practical Use of Prototype Systems*

In solving most artificial intelligence problems, the design of the solution cannot be known in advance. Instead, by doing exploratory programming, using various problem-solving techniques and attempting to produce a prototype solution to a small part of the problem rapidly, you can better understand the problem itself. By continually refining the capabilities of the prototype, the developer evolves a design for a system that ultimately does what was to be accomplished. In the meantime, the successive or "incremental" prototypes provide an additional benefit: the system can be put into actual use far earlier than conventional systems. A carefully constructed expert system, for instance, can already be useful at the 30 to 50 percent completion level. Results from using the system allow users to provide feedback on performance. The feedback is then used to enhance performance, and the process continues until the system performs at the desired level. An additional benefit of fast prototyping is that the system is able to demonstrate its potential usefulness and limitations before a lot of resources have been committed. A project that is clearly not going to produce the required return on investment can be stopped before it consumes a major amount of resources.

Developing the Artificial Intelligence System

The best method for developing a prototype to solve a problem that is amenable to AI techniques is determined both by the structure of the problem and by the available tools. The tools consist basically of various knowledge representation methods and operators, control structures, artificial intelligence languages, integrated utilities (editors, debuggers, code management tools, test managers, etc.), and types of hardware.

In the remainder of this chapter, we will discuss knowledge representation methods, their associated operators, and control structures. Chapter 4 discusses AI languages, Chapter 5 discusses hardware, and Chapter 6 discusses integrated utility sets, among other topics.

- ## Knowledge Representation

Early efforts in artificial intelligence aimed to uncover a small set of powerful reasoning techniques that would solve many problems. Some researchers are still interested in finding principles of this type. In the meantime, people developing applications have opted to find efficient and effective ways to represent large amounts of knowledge to reduce the amount of searching required by a program. These knowledge representation methods are combinations of data structures for storing information and interpretive procedures for making inferences on the stored data. The knowledge representation provides the computer with a description of how the various pieces of information interact.

AI researchers have created a variety of ways to represent different types of knowledge, no single representation being all-encompassing and definitive. The object is to choose a representation that facilitates work on the particular problem at hand. This may extend to combining different representations within a single system or even to developing a new representation or a variation on an existing representation if it suits the problem domain better. The chosen representation structures the problem domain. The problem itself can then be worked out using the representation, various operators, and control strategies.

The knowledge represented in the system plays a variety of roles in solving the problem. It indicates ways to manipulate the features of objects and the relations among them. The knowledge is used to retrieve relevant information, to plot a course of action that has not been explicitly programmed, and to acquire new knowledge.

Artificial intelligence systems may require conventions to represent several kinds of knowledge – knowledge of objects, of relationships, of how events are situated in time, of how to perform actions, and of metaknowledge (knowledge about knowledge). Metaknowledge can provide the system with two ways to fill in missing information: it can supply default values or procedural knowledge that the system can use to work out answers. As an example of metaknowledge, suppose you have a system that keeps inventories in a bookstore by title, and the system has also stored a piece of metaknowledge stating that the only categories of books the store carries are biology, chemistry, physics, mathematics, computer science, and medicine. Suppose someone asks the system whether the store has any cookbooks on French pastry, but can't give the system a particular title. Even though the system can't answer the question by searching through all its titles, it can answer the question on the basis of the metaknowledge.

Different representations can be evaluated along several lines. A good representational system accommodates all the necessary information, makes it easy and efficient to make inferences, suppresses unwanted detail, makes patterns obvious, and facilitates the acquisition of new knowledge and the modification of current knowledge.

Let us now look at several knowledge representation formats – production rules, state-space representations, predicate calculus, frames, scripts, semantic networks, and conceptual dependencies.

- *Production Rules*

    It is often convenient to represent the dynamic nature of an artificial intelligence application by a set of rules made up of conditions and actions. Production rules are a type of "If...then...." rule based on conditions and actions. Descriptions of a given situation or context of a problem are matched to a collection of conditions in a rule that cause the rule's actions to execute, in turn giving rise to new descriptions that "produce" more actions (hence the name "production rule"), and so on until the system either reaches a solution or halts.

    A system consisting of a set of production rules is called a production system or rule-based system. The production rules are the operators in the system; they are what it uses to manipulate the database.

To date, Digital has had more internal experience and success with the rule-based approach to knowledge representation than with other representations methods. This is the method of representation employed in XCON, the expert system that configures orders for VAX supermini-computers and PDP-11 computers.

Production system languages, such as OPS5, facilitate the representation of knowledge in production rules. Each production rule embodies a piece of knowledge that can be added to or deleted from the system, as needed. Both the "If" part and the "then" part of the production rule may be composed of more than one phrase, for instance, "If it's December *and* you're north of the Arctic circle, then conclude that you need a parka *and* conclude that you need a flashlight day and night." The "If...then...." format of production rules makes it relatively easy to code the expert's knowledge because we frequently think about expertise in an "If...then...." manner.

The production rules in expert systems written in OPS5 do not need to adhere to a particular order because the flow of control in an OPS5 program is *not* determined by the order in which the programmer has coded the rules. Instead, rules become candidates to act on the state of the system when their "If" portions are satisfied by the current state of the system, which continually changes as various rules "fire," or act. Explicit control structures in some programs may also limit the applicability of some rules at various times. The capability to add or delete production rules at any point in the system without regard for order helps to make the development job easier.

In all but the smallest systems, it is generally beneficial to split the production rules into sections that perform particular subtasks. This sectioning helps to clarify the order of activation of rule sets and makes maintenance easier.

- *State-Space Representations*
A state is a set of conditions or values that describe a system at a specified point during processing. The state space is the set of all possible states the system could be in during the problem-solving process. To solve a problem using a state-space representation, you move from an initial state in the space to another state, and eventually to a goal state,

by means of operators—typically rules or procedures. A goal is a description of an intended state that has not yet been achieved. The process of solving the problem is the process of finding a sequence of operators, representing a solution path, through the state space to the goal state.

A state-space representation is one in which the operators always produce only one new state in the database each time they are employed. State-space representations, which use various methods of search through the database to achieve solutions, have been used for chess-playing systems, for route-finding systems, and other problems involving many operators and many possible states.

Trees Illustrate Search in State-space Representations The search for a solution path through a state space can be illustrated graphically. Graphs are composed of nodes and connecting links. Each node represents a state of the system, and the links represent the action of an operator to change the system from one state to another. Nodes can contain pointers back to the nodes that preceded them, so when the search is completed, the pathway is preserved. In some search methods, pointers are also used to backtrack if a solution path turns out to be a dead end.

Graphs are generally depicted as either trees or nets. A tree is a graph that begins at the top with a "root" node (tree graphs are generally shown inverted) and whose "descendant nodes" have only one parent node each. This sort of graph has only one path from the root node to any other node. The connecting links between nodes in a tree are called branches. The nodes at the bottom of the tree are called terminal (or leaf) nodes. Nodes in tree graphs are sometimes referred to as parents and children, or descendants.

There is potentially one tree of a state space that represents every possible state of the system and another more limited tree of the path that can be constructed if the program searches efficiently. It can be important to find an efficient way to search because the total potential search space may be infinite or at least so large as to make processing impractical. The classic demonstration of this problem, known as the *combinatorial explosion,* was described by Dr. Claude E. Shannon in an example based on the game of chess.

Combinatorial Explosion In his 1950 paper in *Philosophical Magazine*, "Programming a Computer for Playing Chess," Shannon pointed out that because there are about 30 chess moves that can be made from each position on the playing board, and in the typical game, each player makes an average of 40 moves, there are about $(30^2)^{40}$, or 10^{120}, possible different chess games. Each board position, after each possible move, is a state in the state space. To search a state space this large is obviously a hopeless task.

David L. Waltz, a professor of computer science, Brandeis University, amplified the concept in an article in the October 1982 issue of *Scientific American* entitled "Artificial Intelligence" in which he cites the average number of moves possible at any point in a chess game as 35. He notes that looking ahead only three moves would require a chess-playing system to examine over 1.8 billion moves. So, while we can store knowledge of chess moves, openings, end games, and so forth within a computer, even superfast computers cannot adequately cope with all of the combinations possible in a chess game. Many real-world problems, in fact, demonstrate the combinatorial explosion problem.

The way we limit the searching that an AI application must perform and avoid the combinatorial explosion is to reduce the number of options that must be examined. The name for the tool used to limit search is "heuristic." A heuristic is a guideline, strategy, rule of thumb, or any other method that can improve problem-solving efficiency. While a heuristic doesn't guarantee any solution (let alone an optimal solution), a good one will generally lead to some useful solution in a reasonable amount of time. Heuristics allow the search to be narrowed down to the most reasonable and likely possiblities, making for more efficiency than blind search.

In general, searching involves choosing a path to follow and backtracking to try an alternative if the original path is unsuccessful. Particular methods of search are suited to particular sizes, shapes, and linkages of search trees. The search procedures that are close to blind search, such as depth-first and breadth-first search tend to be more generally usable, while the more focused searches, such as hill-climbing and the A* procedure are generally reserved for a narrower class of problems.

There are several ways to go about informed searching. To avoid backtracking, it is helpful to avoid trying dead-end routes. These dead ends can be highlighted if you expose natural constraints, limiting regularities, in the problem area. A good representation format helps bring out the constraints, making them easier to utilize.

Methods of Searching Depth-First Search: In depth-first search, the order of expansion of the nodes is from the initial node downward in a given path until either the answer is found or the search goes below a predetermined depth bound. If the search exceeds that depth bound or the end of that path, the system backtracks, and search resumes at the nearest node with descendants that have not been expanded. Depth-first search is particularly efficient when there are many long paths that lead to a solution.

Breadth-First Search: In breadth-first search, all the nodes at a given level are expanded before moving down to any of the descendant nodes. If the search must proceed to a descendant node, the search process begins searching the next level from one side and moves horizontally. Breadth-first search helps a system avoid wasting time if there are only a few paths to a solution.

Best-First Search: In best-first search, the systems applies a heuristic function to the nodes at each step in order to choose the one most likely to lead to a solution. This can be very time-consuming; therefore, it is helpful to arbitrarily limit the search by choosing a node above which the search will be abandoned or by setting a value such that as soon as one node is sufficiently promising, the system chooses it.

Hill-Climbing: Hill-climbing begins the same way as depth-first search, but the choice of which of a node's descendants will be expanded is made by estimating which is closer to the goal. This is sufficient when there is some clear way to measure distance from the goal; however, there are deceptive conditions in which the real distance is obscured. Hill-climbing provides a system with ways to overcome these conditions such as backtracking and trying a path that was declined before or applying more than one rule before evaluating the result in order to jump to a different part of the search space.

A Graph of Search in a System
Graphs are illustrations that help us keep track of conditions in a system.

Let us say this graph represents all the states some hypothetical system could take upon the application of various operators. The letters are assigned arbitrarily.

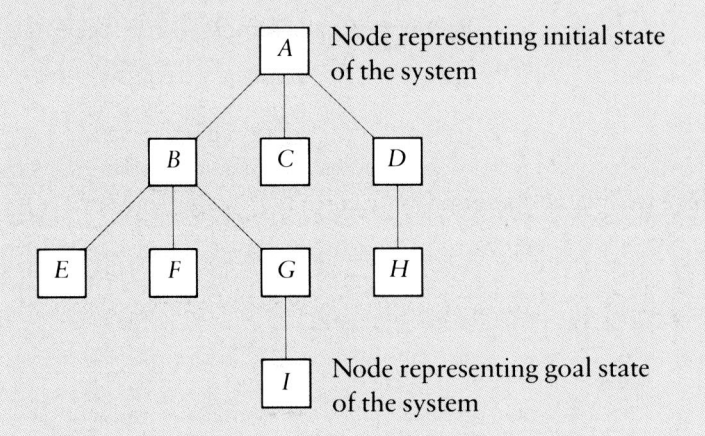

A — Node representing initial state of the system

I — Node representing goal state of the system

Applications of the system's operators may result in 0, 1, or more than 1 new state each time they are applied.

In this example, the most direct solution would be to apply the operators that change the state of the system from *A* to *B* to *G* to *I*. The graph would look like this

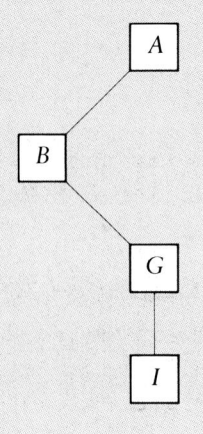

Efficient search methods help us construct this type of limited graph. The more effective the search method, the fewer unproductive nodes are expanded.

Breadth-first Search
This graph illustrates the search path that would be constructed if a system used breadth-first search. Nodes on one level of the graph are expanded before nodes at a deeper level. The path for this search would be *A B C D E F G I*.

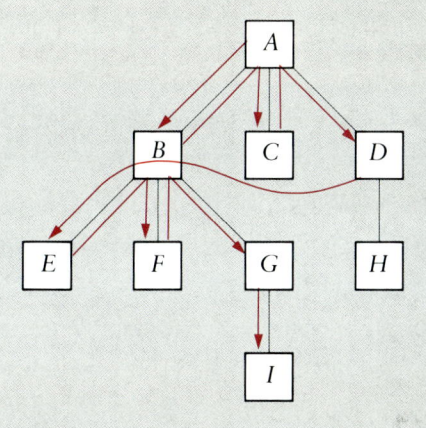

Depth-first Search
In depth-first search, the search proceeds to deeper nodes before expanding nodes on the same level. The path for this search would be *A B E F G I*.

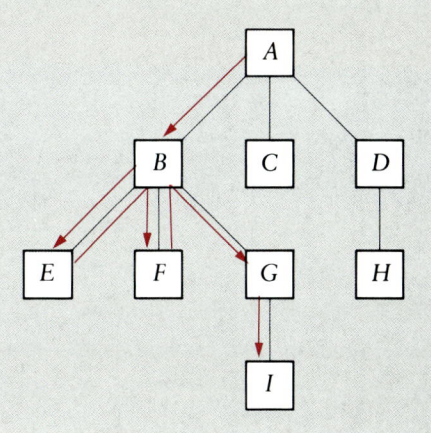

Branch-and-Bound Search: In branch-and bound search, at each step the shortest of the uncompleted paths under consideration is extended one level until the problem is solved. More correctly, search should be terminated when the shortest incomplete path is longer than the shortest complete path. This avoids inadvertently choosing a very long last step that unnecessarily lengthens the path. Because the shortest path was used each time, the first path to reach the goal is necessarily the shortest path overall.

The A* Algorithm Search Procedure: This procedure is a refinement of branch-and-bound search. Rather than choose the shortest partial path to expand, the estimate of distance left to the goal is added to each path before expanding it. The A* algorithm can be used when the remaining distance to the goal is small compared to the total length of the path.

Beam Search: Beam search proceeds horizontally before moving down a level, like breadth-first search, but in beam search, the path moves down only from the best nodes at each level.

Means-Ends Analysis: Some search systems use the differences between the goal state and the current state to guide search. When a difference is identified, a subgoal is generated to eliminate the difference. To achieve that subgoal, other subgoals may first have to be generated and achieved.

- *Predicate Calculus*

Logic, or predicate calculus, like all representation schemes, is simply a way to write expressions about the world. The language of logic has developed over thousands of years and is still actively studied. Using logic, people can make propositions about the world. A proposition is a statement that has a truth value, either true or false.

Connectives are used to combine propositions to form longer statements. The most common connectives correspond to the English words *and, or,* and *not,* and the phrase "*If...then....*"

Individual objects are referred to through the use of constants, variables, or functions. Examples of constants are "Socrates," "ThisBook," and "Hershey'sChocolate." Variables, such as X, Y, and Z, are used to designate as yet unnamed objects. Functions consist of function symbols and the associated object names. A function operates on one or more objects and returns one object. For instance, FAVORITE-FOOD-OF (MICHAEL, PIZZA) could be used to represent the fact that Michael's favorite food is pizza.

We can also make statements about the relations between objects by using predicates. Predicates operate on objects and, unlike functions (which simply return an object when invoked), the value of a predicate is evaluated as either true or false. A string of predicates can be joined by logical connectives to form long propositions.

Rules of inference are rules for deducing new propositions from other propositions already given. There are rules for introducing and eliminating the connectives. A simple rule of inference is called *modus ponens*. This rule lets us deduce a proposition "Q" if we already have the two propositions "If P, then Q" and "P."

When predicates are used as statements, we assume that a true assertion is being made. We can make these assertions by combining constants or variables with predicates. When we make a statement in logic using a variable, we say that the proposition is true for all objects. When we ask a question in logic using a variable, we are asking about the existence of an individual or individuals that make the statement true. To state a rule that all humans are mortal, we use a variable and say:

If human (X) then mortal (X).

To frame the question "Is anyone mortal?" we would use the query:

Mortal (X)?

To solve a problem with logical propositions, we link elements of the problem domain to the function names, predicate names, and constant symbols used in the proposition.

We can use predicate calculus in several ways to do AI problem solving. The first to be developed was theorem proving. To use this method, we describe a domain in logic and reason about it.

After writing propositions about the domain, we need to show that formulas describing the result or query follow from these propositions. The best automatic method today is still resolution. This technique is basically a proof by contradiction. It assumes the result is false and shows that it leads to a contradiction.

Resolution is nice in that there is only one rule of inference. The program doesn't have to be concerned with which rule to pick. However, resolution is slow and the proofs are hard for a person to follow. Natural deduction theorem-proving systems use inference rules that people can follow better. Domain specific heuristics are often included to direct the proof. This technique is still an area of AI research.

We do not yet understand the computational properties of logic. That is, we don't understand how to treat logical propositions as programs. Recent attempts, such as the development of the Prolog language, have succeeded only partially. There are many who believe that any representation system must have a basis in logic. What remains to be solved is how to control deductions.

- *Frames*

Frames and an associated knowledge representation form called the script organize knowledge in ways that make it easy to understand how inferences can be made. Professor Marvin Minsky of MIT developed the frame representation. Professor Roger C. Schank of Yale developed the script representation and has made use of it in natural language understanding programs.

Frames, as in "frames of reference," cluster closely associated knowledge about objects and events. Frames help a system interpret meaning according to context and provide it with a store of associated detail from which to infer missing elements. Typically, a frame describes a class of objects such as *bicycle* or *sports*. The frame consists of a collection of slots that describe aspects of a situation, action, object, or event. These slots are filled by values, procedures, or even by pointers to other frames. A frame for a bicycle might look like this:

Frame: *William's-bicycle*
Purpose: *recreation*
Wheels: *two*
Source of Power: *muscle*
Brand Name: *Fuji*

A set of conditions that must be met by the filler, and/or a default value when none is known, can be associated with each slot. For instance, in the *William's-bicycle* frame, we could have specified

Brand Name: a PROPER NAME (DEFAULT = *Schwinn*)

Procedures may be associated with particular slots. For example, it is often useful to tell the system to carry out some action whenever a slot is filled or to tell the system how a value for a slot can be computed if it is required. One frame or one element in a frame slot can point to another frame or slot, which can point to another, and so on, allowing for extensive inferencing.

In order for a system to solve a problem using this representation, we need to write a program that uses the frames as data, as in a database system. The artificial intelligence system can then make inferences by filling in the linking or inheritance relationships associating one slot or frame with another slot or frame.

- *Scripts*

A script is a framelike structure that uses stereotyped situations to represent knowledge in a particular context. Scripts are unique in having a time orientation: they are concerned with sequencing cause and effect. The script structure consists of a set of slots that correspond to a set of aspects of an event that would occur in the typical situation. If no specific information is available with which to fill a particular slot, the system fills the slot with a default value supplied earlier by the programmer for just such occasions. If events in a situation analyzed by a script system go according to the stereotype, the system can then predict by analogy that other events will occur. However, if an unexpected event occurs, the system recognizes that the script no longer applies, and no further predictions can be made.

A script for a birthday party might include the presentation of a cake, singing the birthday song, blowing out candles, and opening presents. A system supplied with a script can make use of expectations about situations in interpreting input. Suppose that a natural language system with a birthday script were given the input sentence, "June and Bill gave Graham a cake with candles, and he blew them out." The script, equipped with knowledge about the birthday candle ritual, would enable the system to assign the ambiguous pronoun references in the text "he blew them out" to the appropriate referents.

- *Semantic Networks*

Semantic networks can be illustrated by diagrams consisting of nodes and arcs. The nodes represent objects, actions, or events. The connecting arcs, called links, represent the relations among the nodes. A link could mean that the object on one end is an attribute of an object on the other end, or it could mean that one implies the other, or anything else you define the link to mean. This information is stored in the computer as attribute-value memory structures or property lists, depending on what language is used to implement the system.

The following example shows how semantic nets can facilitate the inheritance of values. In the semantic net below, we have a link representing the relationship *is-a* pointing from a node representing *swimming* to a node representing *aerobic-sport*. The link representing *is-a*, pointing from *Australian crawl* to *swimming*, allows the system to infer that the Australian crawl is a subset of *swimming*, which is a subset of *aerobic-sport*, so that the Australian crawl is an aerobic sport.

$$\text{AUSTRALIAN CRAWL} \xrightarrow{is\text{-}a} \text{SWIMMING} \xrightarrow{is\text{-}a} \text{AEROBIC SPORT}$$

The transitivity that makes inferences like "Australian crawl *is-a* aerobic-sport" is also the danger in semantic nets. It is important to be aware of what the link represents or a special case may end up falsely describing members of a more general class.

More than one net can apply to a single object in a representation. These different nets, spoken of as "perspectives," signify the different contexts in which the object can be described. For instance, another perspective on the Australian crawl could be devoted to races and records.

- *Conceptual Dependencies*

Conceptual dependencies are representations used to store information in natural language systems. A conceptual dependency maps the meaning of sentences to directed graphs composed of a small number of primitives, or basic concepts, signifying actions, states, or changes of state. A system can combine these primitives as needed to form complex statements. A natural language system paraphrases the underlying concepts of input text, putting them in a form the system can use in drawing inferences, performing translations, or answering questions by referring to a database. The paraphrase is an intermediate language written in graphs of nodes and arrows and suited to computer manipulation.

Conceptual dependencies equipped with eleven primitives have been utilized by Professor Roger C. Schank. Among the primitives Schank has used are *atrans*, which signifies the change in possession of an object; *ingest*, the taking in of food, liquid, or gas by an animal; and *mbuild*, which is the construction by an animal of new information from old information.

Schank uses the primitives along with a few other concept types to generate the paraphrases of sentences.

Conceptual dependencies can be used to translate text from one language to another, with the conceptual dependency paraphrase mediating between the two natural languages. Dr. Yorick Wilks, for example, used a similar method to develop an English-to-French translation system. Wilks's system used sixty semantic primitives in classes that described entities and actions.

The use of primitives allows a system to represent the similarities and differences in sentences whose deep meanings may be less obvious when they are fleshed out in English. The analysis into primitives also provides a basis for making inferences.

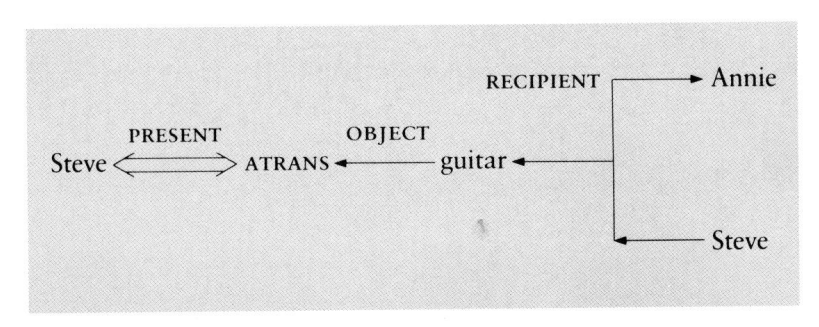

This represents the atrans "Steve gives Annie a guitar."

Primitives make inferencing more efficient because each primitive carries a set of implications with it in the system. For instance, the primitive *atrans,* which means an abstract transfer of the possession of something, implies that one person used to own something and now does not, another person now owns something who did not own it before, and a transfer of that thing has occurred from the former owner to the present owner.

Other Techniques and Devices Used in AI Systems

Now that we have surveyed some basic knowledge representation forms, let's take a look at some of the techniques that may be applied to various AI systems.

Abstraction

Abstraction methods are used to sift out the important features from among many details. This is particularly useful in visual and natural language systems where too much detail can make the representation unintelligible. The abstraction process is also used in other applications. Solving a simplified version of a problem frequently sheds light on how you might solve a more complex but analogous problem.

Inheritance

Inheritance is the capability that allows a system to pass values through link or slot relations. It can provide slots with default values, and it increases programming efficiency. Inheritance simplifies programming because some relations can be left implicit instead of written out in rules.

There is an expectation in systems structured to take advantage of inheritance (such as frames or scripts) that a value in a descendant node will inherit the characteristics of a parent node. The following frames provide an example:

Frame: *Duck*
Distinguishing Feature: *webbed-feet*
Preferred mode of locomotion: *swimming*
Habitat: *lakes*
Diet: *bugs-and-plants*

Frame: *Christine's-duck*
Is-a: *duck*
Color: *yellow*
Name: *Skinny*
Habitat: *swimming-pool*

If we wanted the system to make use of inheritance, we would set up the *Christine's-duck* frame to inherit the values of the more general *Duck* frame through the Is-a relation. However, because Christine's duck no longer lives in the wild, we have prevented the automatic inheritance of the Habitat value by specifying a Habitat slot to override the one from the *Duck* frame. Inheritance lets the system know about Skinny, in addition to the facts that he's yellow and lives in a swimming pool, that he has webbed feet, prefers to get around by swimming, and eats bugs and plants.

Passing along values by inheritance adds efficiency to the representation because it is not necessary to repeat and make explicit all relevant aspects of each object described in the system. If we also wanted to describe the ducks at Lake Ellyn, Odette's ducks, and the ducks at Brookfield Zoo, we could make use of the information in the *Duck* frame without having to encode it separately for each instance.

- *Grammars and Parsing*

Parsing is used in the natural language area of AI to relate meaning to the structure of verbal representations. Parsing, in AI applications, is the conversion of syntax to a form usable by the computer. That is, parsing breaks a sentence down into components the system can identify as various parts of speech with specialized functions. The computer can then construct a representation, called a parse tree, that portrays the relationships among the parts of the sentence and can be used to draw inferences.

Parsers use grammars to identify the classes into which words in the input sentence can be assigned. Grammars formalize the allowable relationships of those classes. The classes of words might be subjects and verbs and other traditional parts of speech, or they might be other categories the grammar's designer defines. As an input sentence enters the parsing component of a system, the parser uses the grammar to classify the words. Then the parser tries to identify the rule or rules that are appropriate to build the tree that shows the relationships of all of the words. To build that tree is to parse the sentence. Meaning is assigned to the parsed sentence by matching information contained in the database.

- *Augmented Transition Networks*

Augmented transition networks (ATNs), invented by Dr. William Woods, are one of the most widely used mechanisms for parsing sentences in natural-language systems. ATNs are networks of nodes and arcs that collectively represent the syntax of a language. The nodes in the ATNs have syntactic labels, like *noun, preposition,* or *verb phrase.* The arcs are labeled by vocabulary words. There can be ATNs for whole sentences as well as for phrases and clauses, and the clause and phrase ATNs may be called as subroutines during the processing of larger units, like sentences. As a piece of input text is fed into an ATN, its elements must match the parts of speech and come in the same order as the nodes in the string. If the system is able to match the input to an ATN string, the sentence or phrase is considered to be "recognized" by the system. The system can then store the sentence, labeling its elements in a way that aids interpretation in a natural language system.

- *Chaining*

Motion, or "chaining," along paths to be searched in a state space, or in the firing of production rules in an expert system, can go forward or backward. Some systems use both forward and backward chaining.

Forward chaining, also called bottom-up processing, antecedent reasoning, or data-driven strategy, proceeds by manipulating data or antecedents in the knowledge base to produce a modified status of the system. This kind of processing is useful in "constructive" systems that aim for synthesis, as in manufacturing or in examining what-if questions.

Backward chaining, also called top-down processing, consequent reasoning, or goal-directed strategy, breaks down the goal statement into subgoals and works backward to see if the necessary and sufficient antecedents that would satisfy the goal are present in the database.

Some rule-based, or production, systems can run either forward or backward. In forward chaining, the systems work from known "Ifs" to newly deduced "thens." Backward chaining systems hypothesize a conclusion and work backward from "thens" to find antecedent "Ifs" that support them.

Using either direction of chaining, however, the system may make use of backtracking. Backtracking means that when the system reaches an obstacle, such as a dead end, in the search, it backs up and tries to follow a different branch.

- *Explanation Facilities*

Because many AI systems follow paths through complicated territory—exploring many states, searching, using many rules, etc.—and it is not immediately obvious to the user just how the system reasoned its way through, it is helpful to arrange for the system to explain or justify its reasoning. Then the user can evaluate whether the system is working reliably or needs some revision. This is an important feature of many expert systems. Most explanations consist of listing the steps taken through the problem space, and many include the reason for taking those steps.

- *Control Strategies*

Control strategies are used to systematically activate procedures in the system. Procedures have various purposes. Some perform operations on the data, some work out a method to perform operations, and some show that neither of the former two objectives can be met.

Procedures themselves sometimes control other procedures; that is, a procedure can invoke the subprocedures it needs. Or control can be attached to objects that hold pointers to appropriate procedures to use with those objects. Control may also be handled by allowing procedures to "volunteer" themselves for use in response to a request in the system, the most appropriate one being chosen.

When a system brings a number of knowledge sources to bear on a problem, a need may arise for those processes to share information. For instance, in a speech understanding system that utilizes knowledge of phonetics, semantics, and the rules of conversation, hypotheses about meaning generated by one source can be posted and checked by the others, and the system can look for evidence to confirm an interpretation.

One of the ways for procedures to communicate among themselves is the blackboard method. A "blackboard" in the system acts as a central focus where procedures can leave instructions for, and receive results from, other procedures. The blackboard method can be specialized so that either each procedure has a special spot reserved for the messages it sends and receives or so that there are special spots devoted to different kinds of tasks, which any procedure can use.

Control devices such as blackboards come in handy in planning. Typically, planning breaks a problem down into subproblems that are easier to solve than the original problem. The subproblem solutions are then combined to produce a solution to the original problem. Devices like blackboards are helpful for handling interactions among the subproblems during processing.

Another way for procedures to talk to each other is to send messages. A procedure might make a request or announce that some action has been accomplished. In object-oriented programming, data objects store their own interpretations of directions. You can send several different objects the same command message, and they will each carry it out in the way tailored specifically for them.

"Demons" are procedures designed to respond to specific situations that arise during processing. They "watch" registers, indexes, values, or blackboards for indications that they are needed, at which time they are activated. If more than one demon is activated at one time, priority is decided by the system's rules.

- *Confidence Factors*

Confidence, or certainty, factors are used with an expert system to provide an index of how strongly the system's knowledge supports its final, cumulative conclusion. Confidence factors are usually expressed as decimal fractions between 0 and 1.0, inclusive, that rate how definite a piece of information used in the system is.

If you are using an expert system to predict the weather in Boston, you might assign a confidence factor of 0.8 to the statement that if it is humid in early August and a Bermuda high sets in, then it will rain. When the system uses this statement and combines this factor with other factors that bear on rain in early August, it can come up with an overall confidence factor.

In real systems, the factors are combined by means of some ad hoc application of probability theory. Often the way confidence factors are derived and worse yet—propagated—in systems is about as reliable as long-range weather predictions. Nevertheless, where credible confidence factors are assigned and a reasonable, well-understood algorithm is used to apply them, these factors can be used to help guide the search process as well as to evaluate the strength of the conclusion reached by an AI system.

The important question to ask in looking at any system's use of confidence factors or probabilities is how sensitive its problem solving ability is to the values chosen.

Choosing a Language

Once the developer has chosen a form of representation for the problem domain, he or she can implement the system in the language that best facilitates those goals.

In addition to specific AI languages, such as OPS5, and languages like LISP and Prolog that are more like traditional languages, it is possible to use some general purpose languages such as C, FORTRAN, and PL/1 to program some AI systems.

Conventional languages also are frequently helpful for developing portions of AI systems that are written, in the main, in an AI language. For instance, a system might need database techniques and statistical processing. The system becomes more flexible and easier to develop if the chosen hardware and software allow you to intermix AI and conventional processing. As an example, Digital's VAX LISP and VAX OPS5 allow you to make use of conventional computing tools as well as AI tools. With VAX LISP, you can call out to FORTRAN, for instance. This allows the use of previously written software, thus avoiding the need to convert existing programs to an AI language. With VAX OPS5, you can both call out and call in, so FORTRAN can do data collection and OPS5 can make the inferences. Another example is that in time-critical applications, you need to provide a response in a predictable time period. Several conventional languages provide excellent realtime facilities. A system that requires both realtime and intelligent inferencing is most efficiently built from a combination of tools and languages.

In the next chapter, we will examine the programming languages used most often in artificial intelligence applications.

✺ Annotated Bibliography

Barr, Avron, and Edward A. Feigenbaum (eds.), Vol. I and II and Paul R. Cohen and Edward A. Feigenbaum (eds.), Vol. III, *The Handbook of Artificial Intelligence*, 3 Volumes, Stanford, California, HeurisTech Press, Vol. I, 1981, Vols. II and III, 1982. © by William Kaufman, Inc., Los Altos, California, 1981. This reference set discusses, among other topics, various logics, knowledge bases, search, graph representations, planning and problem solving, and problem reduction.

Lenat, Douglas, B., "Computer Software for Intelligent Systems," *Scientific American*, September 1984, pp. 204-213. This article discusses the power of knowledge to constrain search in AI systems, inferencing methods, and applications of AI techniques to expert systems.

Niwa, Kiyoshi, et al., "An Experimental Comparison of Knowledge Representation Schemes," *AI Magazine*, summer 1984, pp. 29-36. In this article, Niwa et al. compare four knowledge representation schemes applied to one expert system problem.

Rauch-Hindin, Wendy, "Artificial Intelligence: A Solution Whose Time Has Come," *Systems & Software*, December 1983, pp. 150-177. This article discusses AI applications, expert systems mechanisms, knowledge engineering tools, and AI languages.

Rich, Elaine, *Artificial Intelligence*, New York, McGraw-Hill Book Company, 1983. This textbook discusses knowledge representation methods, logic and reasoning, control strategies, heuristic search, and problem-solving techniques.

Shannon, Claude E., "Programming a Computer for Playing Chess," *Philosophical Magazine*, Series 7, Vol. 41, 1950, pp. 256-275. In this paper Shannon discusses the combinatorial explosion problem in the context of programming a computer for playing chess.

Waldrop, M. Mitchell, "The Necessity of Knowledge," *Science*, Vol. 223, March 23, 1984, pp. 1279-1282. In the article, Waldrop discusses the idea that the essence of intelligence seems to be less a matter of reasoning ability than of knowing a lot about the world.

Waltz, David, "Artificial Intelligence," *Scientific American,* October 1982, pp. 118-133. This article discusses heuristic search, planning and control strategy, computer learning, constraint propagation, vision and language processing.

Winston, Patrick Henry, *Artificial Intelligence* (2nd ed.), Reading, Massachusetts, Addison-Wesley Publishing Company, 1984. Winston's textbook discusses knowledge representations, logic and theorem proving, program control strategies, rule-based systems, exploiting constraints, semantic nets, search, nets and trees, problem reduction, and state spaces, among other topics.

Zadeh, Lofti A., "Making Computers Think Like People," *IEEE Spectrum.* August 1984, pp. 26-32. In this article, Zadeh discusses the concepts of "fuzzy logic" and fuzzy sets."

Chapter 4

**The Languages
of Artificial Intelligence**

With conventional data processing, we solve problems by computing values and assigning them to variables. With symbolic processing, we represent objects from the real world and manipulate the relationships between those objects. An example of symbolic processing is word processing. A word processor manipulates objects like letters, words, and paragraphs by such actions as "cutting and pasting," capitalizing, and adding and deleting. In terms of the activity of word processing, the value of the word is less important than how the letters, characters, and words relate to their settings. To state it differently, while the word processing system does have to compute values and assign them to variables, computing those values is only a process within the larger goal of manipulating the relationships of the objects within their setting. And you can more readily represent these relationships in a symbolic language than as numbers, values, or variables in conventional languages.

Because artificial intelligence programs are geared primarily to symbolic processing rather than to numeric computing, it might well be expected that programming languages for AI work would have characteristics different from those of languages suited to traditional data processing. While it is possible to do artificial intelligence work with data processing languages, artificial intelligence researchers have developed special languages that make AI programming much easier.

No one language is ideal for all AI applications or for all AI programmers. Features that process some applications efficiently are not only irrelevant to other applications but are directly counterproductive to others. And programmers vary in the criteria they set for an optimal language, depending, among other things, on their personal working styles. Some programmers prefer a low-level language that gives them the ability to build in their own efficiency-optimizing features, while other programmers prefer to take advantage of the built-in facilities of full-fledged programming environments, including good editors, interactive debugging facilities, and input/output routines.

Among the features that different programmers consider desirable in AI languages are the ability to manipulate lists; the ability to accommodate a variety of data types; late binding times for the size of data structures and for object typing; pattern-matching; an interactive mode of operation; and inference methods that can be programmed to automatically make decisions and deductions and that can store these results as needed.[1]

The leading AI languages today are LISP and Prolog. In this chapter, we will examine these and a few other languages currently important in AI.

Some languages formerly used in AI that deserve mention but are now considered obsolete are

- IPL-II, the first list processing language was developed by Allen Newell (now University Professor of Computer Science at CMU) Nobel laureate Herbert Simon (now Richard King Mellon Professor of Computer Science at CMU) and J.C. Shaw (now a computing software consultant) for a chess program.
- SAIL was derived from ALGOL. SAIL supports an associative memory, which is helpful in accessing facts associated with more than one index, and it has features that are useful for conventional computing. SAIL was designed by researchers at the Stanford University AI Laboratory for vision and speech-understanding systems.
- CONNIVER, written by Gerald Sussman (a professor of electrical engineering at MIT), was developed from an earlier language called PLANNER and offers the programmer more flexible control. CONNIVER also originated the use of a tree of contexts in the database to explicitly represent control information—much as a simple stack encodes control information in an ALGOL-like language.
- KRL, a frame-based language written by Dr. Daniel G. Bobrow (now at Xerox Palo Alto Research Center) and Terry Winograd (an associate professor of computing science at Stanford University), employs pattern-matching and does resource-limited reasoning using an agenda.

Each area of AI has its own specialized languages, of which there are hundreds. The following are specialized languages used in robotics but not in the other areas of AI:

- Machine Intelligence BASIC, an enhanced version of BASIC, created by Machine Intelligence Corporation, Sunnyvale, California, for robot functions.[2]
- RAIL® (Robot Automation, Inc., Language), developed by Automatix, Inc., Billerica, Massachusetts. RAIL was designed to accommodate vision applications as well as more general programming functions.[2]
- AL, developed by students of Professor Thomas Binford at Stanford University's Artificial Intelligence Laboratory. AL was designed to compute and keep track of robot movements and positions in space.[2]
- VAL,† from Unimation, Inc., a Westinghouse company, was introduced in 1979. VAL's interpretive, interactive qualities lend themselves to practical applications.[2] VAL is a computer-based control system and programming language designed for use in Unimation, Inc. industrial robots.
- AML (A Manufacturing Language), developed by International Business Machines Corporation, is a high-level interactive programming language. AML was designed and written for manufacturing engineers and engineering applications programmers. Features include sensing and intelligence—in this case, the ability to use information to modify system behavior in preprogrammed ways.

LISP

AI is considered a young field, but LISP, its premier programming language, is, relatively speaking, an old-timer. The only older programming language still in use is FORTRAN. While FORTRAN was designed primarily for numerical computation, LISP was designed primarily for manipulating symbols. Symbolic processing languages, such as LISP, extend our ability to use computers from the relatively smaller realm of numeric problems to the larger realm in which we work in words and symbols.

LISP, developed by John McCarthy (now professor of computer science at Stanford University) at the Massachusetts Institute of Technology in the late 1950s, stands for LISt Processor. LISP programs consist of collections of procedures in list form that operate together to accomplish a given purpose.

A LISP list is a string of "atoms," the basic elements that the system will manipulate, enclosed by parentheses. A list can be empty or can consist of either atoms (such as numbers, symbols, or words) or other lists.

LISP lists are enclosed in parentheses. Procedure calls in LISP are written as lists, consisting of an open parenthesis, a procedure name or symbol, such as "EQUAL" or " + " that tells what operation is to be performed; the arguments, or symbols, upon which the procedure is to operate; and finally, a close parenthesis. For example, (+ 2 3) is a LISP procedure. A single quotation mark precedes an expression that the LISP processor should treat as data rather than as a procedure: '(+ 2 3) indicates a list of three atoms, rather than an addition procedure.

LISP programs consist of procedures you select to work together. LISP comes with a variety of ready-made procedures, like " + ," called functions. Some procedures are called functions. Functions operate on the arguments following them within the parentheses in order to produce values. Most modern LISPs come supplied with several hundred functions and, as with procedures in general, you can define your own. A function returns a value that can be used immediately where it's called. If you were to type (+ 2 3) into a LISP language processor, the system would evaluate the expression and return the response of 5. The value returned by the system as a result of evaluating an expression could be any kind of LISP object: a number, a data structure, or even a program. When you use a procedure to compute the value of, or to evaluate, an expression, LISP responds with, or returns, a value.

The arguments, or symbols, to which a LISP procedure name or symbol refers may be constants, variables, or even other lists. LISP variables may represent anything, not just a number. LISP also supplies many specific data types, such as integer, character, string, and structure.

While LISP comes supplied with many basic procedures such as mathematical functions, predicates (expressions that evaluate to true or false), and logical connectives, LISP is best characterized by its list manipulation procedures, CAR and CDR. CAR directs the processor to return the first item in a list, while CDR (pronounced "could-er") returns all but the first:

(CAR '(PINE PALM OAK)) ;returns PINE
(CDR '(PINE PALM OAK)) ;returns (PALM OAK)

These two operators are all you need to search through the binary tree data structure formed by all LISP programs. In addition to commands for searching the tree, LISP gives you commands that build the data structure, such as CONS, which adds elements to lists. These lists can be modified and linked together to build large and complex data structures. LISP data structures can accommodate a variety of knowledge representations, such as predicate logic, "If...then...." rules, and frame representations.

The ease and power of recursion in LISP programs are notable. When you have solved a portion of a problem, and the problem-solving method is applicable to the remaining portion of the problem, LISP allows you to define a function to use itself repeatedly on subproblems.

LISP is a flexible language that you can modify for your own needs. You can write code ranging from operating systems to high-level programs in LISP. In fact, LISP itself can be written in LISP. LISP makes no distinction between lists that contain data and lists that contain programs. This makes it easy for LISP programs to manipulate or even generate other LISP programs. In addition, it is possible to integrate data and information about procedures. This integration forms the basis for sophisticated "frame" and "object-based" systems in common use in AI applications.

All LISPs come with powerful program development facilities, like program debuggers, editors, and tracing and stepping facilities. In addition, the fast, interactive way LISP returns a value immediately when you type in a procedure is helpful when you're doing exploratory programming. LISP can also be compiled to run faster—in most applications the compiled code is much more compact and runs at least ten times faster than interpreted code.

LISP has become a mature, sophisticated language. After over two decades of use and development, a standard has emerged. In fact, the U.S. Defense Advanced Research Projects Agency has specified COMMON LISP as its recommended standard for all AI-related work. Dr. Guy L. Steele, Jr., currently with Thinking Machines, Inc. of Cambridge, Massachusetts, is the author of a book describing the currently agreed upon definition, titled *COMMON LISP: the Language,* Digital Press, Maynard, Massachusetts, 1984. This standardization of the language was developed in cooperation with many individuals from universities, research laboratories, and the commercial computer industry. The need for a standard LISP was felt because, over the years, users had taken advantage of LISP's flexibility, changing it for their own purposes so much that they usually could not share code. The wide variations in LISP dialects prevented the development of commercial implementations. Now that LISP users have united to standardize and stabilize COMMON LISP, commercial implementations are possible.

COMMON LISP is a successor to ZETALISP, but it pays its respects to MACLISP and INTERLISP dialects as well. COMMON LISP provides a base dialect from which can stem implementations for personal computers, commercial timeshared computers, and supercomputers. For this reason, COMMON LISP omits features that are useful only on some classes of processors.

Digital's VAX LISP is a fully supported implementation of COMMON LISP that provides an interpreter, a compiler, a debugger, a *pretty printer* (which formats printed output for readability), and a powerful user-extensible text editor. A subset of COMMON LISP, GCLISP† (Golden Common LISP), developed by Gold Hill Computers, Inc., Cambridge, Massachusetts, and available from Digital Equipment Corporation, is also available for Digital's Rainbow personal computer and the IBM PC. COMMON LISP is also supported on Symbolics,† Inc., LISP machines.

Other dialects of LISP include

- MACLISP, which was developed at the Massachusetts Institute of Technology to run on PDP-10 and -20 (DECsystem-10 and -20) computers. MACLISP emphasizes speed, economy of storage, and the flexibility to custom build your own tools. MACLISP has many enthusiasts in the research and development community. This dialect is the predecessor of "LISP machine" LISP. MACSYMA,† a program used widely in mathematical and physics research environments, was developed in MACLISP.
- INTERLISP was developed by Bolt Beranek and Newman, Inc., Cambridge, Massachusetts, and Xerox Palo Alto Research Center, Palo Alto, California. The original implementation was on the PDP-10 computer. It was intended for use in exploratory programming. The emphasis in this dialect is not on speed or economy of memory but on a

helpful programming environment. Only INTERLISP has DWIM (Do What I Mean), a spelling correction facility that allows the interpreter to execute what it "assumes" the user meant to type. INTERLISP also has MASTERSCOPE, a database that works like an online interactive cross-reference tool to aid in debugging and program development. MASTERSCOPE finds all the places a function is called. INTERLISP is better supported and documented than MACLISP and has been better controlled to maintain compatibility of older programs with newer implementations. However, there is no truly standard INTERLISP.

- INTERLISP for VAX, available from Digital Equipment Corporation, runs on VAX superminicomputers under the VMS and UNIX† BSD 4.2 operating systems. Both implementations were developed by Information Sciences Institute at the University of Southern California.
- Franz LISP, developed at the University of California at Berkeley, was intended to be MACLISP for VAX computers, allowing them to run MACSYMA. This implementation is available on the UNIX operating system on VAX computers. Franz LISP is currently distributed and supported by Franz Inc., Berkeley, California.
- ZETALISP, an enhanced version of the Massachusetts Institute of Technology's LISP machine LISP, runs on some LISP machines. ZETALISP supports extensible data structures and object-oriented programming and has a highly sophisticated interactive development environment including online ZETALISP documentation and a collection of programming facilities providing all the functionality of MASTERSCOPE and more.
- PSL (Portable Standard LISP), implemented by the University of Utah, is a transportable LISP that is gaining some acceptance. One of PSL's strongest features is that transportation of software between machines is designed to be completely transparent.
- PSL is available on Digital's DECSYSTEM-20 and VAX computers.
- DECSYSTEM-20 Common LISP (CLISP), implemented by Rutgers University.

Prolog

Prolog, which was named for *programming in logic*, was developed at the University of Marseilles by Professor Alain Colmerauer and his colleagues in the early 1970s. Dr. David H. D. Warren (now vice-president of engineering, Quintus Computer Systems, Inc., Palo Alto,

California), when at the University of Edinburgh in Scotland, created an implementation of Prolog for the DECsystem-10 computer that included both an interpreter and a compiler. This implementation was called DEC-10/20 Prolog or Prolog-20. Other centers of Prolog development have been London and Budapest.

Like LISP, Prolog was designed for the manipulation of symbols, and both languages lend themselves to expressing predicate calculus logic. Prolog is also interactive, like LISP. Prolog, however, is characterized as a relation-processor rather than as a list-processor like LISP. Prolog was intended for use in natural language processing systems, but has also shown its usefulness in the areas of computer-aided architectural design, expert systems, and database building and query systems.

Prolog is designed in such a way as to automate search through a tree-structured domain or knowledge base. Since many ATNs and semantic nets have treelike shapes, Prolog has naturally been applied to language and query applications.

Prolog shows some promise as a suitable language for parallel processing systems now beginning to be developed. In a massively parallel processor (MPP), each node in the tree structure would be assigned to a separate processor. A Prolog application would propagate through the MPP by passing messages to activate links to nodes. In a parallel processor with only a few nodes, a Prolog program could assign tasks to processors each time the program reaches a branch point.

While LISP has been the language of choice for artificial intelligence in the United States, Prolog has been the leader in Japan, France, the United Kingdom, and Hungary. One reason may be that Prolog programs are smaller and are easier to read than equivalent LISP programs. Another reason is that Prolog's logic-based semantics hold the promise of helping to simplify the representation of knowledge. If Prolog is an inherently parallel-processing language as many contend, then it is a good fit for the parallel processing computers the Japanese hope to build in their Fifth-Generation Computer Project.

- *Prolog Notation*

A Prolog program consists of clauses (implications). Each clause has no more than one consequent conclusion. The antecedents are not restricted, and may consist of any number of conclusions joined by the logical connectives "and" and "or." If no antecedent is stated, then the clause is a simple fact. If one or more antecedents is stated, the clause is called a rule.

Prolog clauses are written with the conclusion first, followed by a backwards arrow,‡ followed by the antecedent(s). As an example:

married-to(martin,ellen) :-husband-of(martin,ellen).

is read as "Martin is married to Ellen implies that (or "provided that") Martin is the husband of Ellen."

Such a declarative statement can also be used as the basis of a proof. To prove that Martin is married to Ellen, prove that Martin is the husband of Ellen. These proof procedures are similar to the "If...then...." rules of production systems. It is the role of the Prolog interpreter to try to prove queries by comparing them to statements in the database. Programs are executed by logical deduction from the clauses.

Prolog statements are a simple, straightforward way to specify data and relationships. Coding is made even easier when you have a truth that can be generalized. In that case, instead of coding each instance in which some fact is true, you can use variables to code a rule that covers all the instances. (Variables are capital letters or begin with capital letters.) A more general version of the rule presented earlier would be

married-to(X,Y) :- husband-of(X,Y).

Prolog notation puts the predicate (a word that describes a relationship) first followed by the arguments (the subjects that are related) and encloses them in parentheses.

A Prolog database is made up of statements and rules such as

likes(bonnie,flowers).
likes(bonnie,dolphins).
likes(bonnie,saxophones).

that say that Bonnie likes flowers, Bonnie likes dolphins, and Bonnie likes saxophones.

---

‡ Since most computer keyboards don't have a backwards arrow character, :- is most commonly substituted.

To create a generalized rule, we could write:

likes(bonnie,X) :- is-a(X,flower).

This rule says that Bonnie likes X provided that X is-a flower. Rules allow the system to derive statements from other information in the database. If this database were supplied with a list that identified certain objects as flowers, such as peonies, roses, and tulips, the system would be able to use the rule to derive the information that Bonnie likes specific objects known as peonies, roses, and tulips.

Prolog has powerful pattern-matching ability. A question in Prolog is considered a goal (see Chapter 3 for more information on goals). When the interpreter is presented with the clause to the right of the query symbol, ?-, it searches the database for a match. If Prolog finds a match, it returns the appropriate argument or "yes." If it doesn't find a match, it returns "no." In this example, we could ask the system if Bonnie likes flowers:

?-likes(bonnie, flowers).

And the system would return

yes

Or we could ask more generally if Bonnie likes anything:

?-likes(bonnie,X).

Prolog would return

X = flowers

If we wanted more responses, we could type a semicolon and a return after a value Prolog returned, and Prolog would find another true response if the system contained one.

X = flowers;
X = dolphins;
X = saxophones

If you simply type a return, Prolog concludes that you are satisfied with the response and looks no further.

Prolog does depth-first search with backtracking. In other words, the program follows a particular line of inferences until reaching a statement that fails to lead to a solution, at which time, the program backtracks to a point that leads down some other potential solution path. The drawback to this format is that it makes the program more susceptible to the combinatorial explosion problem. This problem can be avoided if you structure the program's queries so that the more specific searches are conducted before undertaking the general searches. Other than this consideration, the statements in a Prolog program can be in any order because each can be evaluated on its own as a fact.

- *Prolog for Digital Equipment Corporation Computers*
A variety of Prolog implementations is available. Quintus Prolog, developed by Quintus Computer Systems, Inc., Palo Alto, California, runs on VAX computers under the VMS and UNIX operating systems, as well as on other machines.

In Europe, Digital Equipment Corporation offers Prolog II, developed by Prologia, of Marseilles. This arithmetic implementation of the Marseilles Prolog is a new version of the language that runs on both VAX computers and the Rainbow personal computer. Prolog II has advanced features like infinite trees, which are useful when you want to specify an infinite recursive structure but you want the language implementation to build only the part of the tree you are currently examining. Prolog II is designed to be portable.

MPROLOG, implemented by SZKI in Hungary and marketed internationally by Logicware, Inc., Toronto, Canada, runs on VAX superminicomputers under the VMS and UNIX operating systems.

There is also a DECsystem-10/DECSYSTEM-20 Prolog implementation designed specifically for those computers.

C-Prolog (implemented in the C language), developed by Dr. Fernando C. N. Pereira, runs on VAX superminicomputers and other machines. It is considered to be fairly portable. The C-Prolog interpreter supports the Edinburgh syntax and semantics and is source compatible with DEC-10 Prolog, considered to be the language standard.

OPS5: A Language for Building Production Systems

The OPS language was created by Dr. Charles L. Forgy, a research computer scientist at Carnegie-Mellon University, in the late 1970s for building large, forward-or backward-chaining, production-based expert systems. Because of this emphasis, OPS5 is not considered a general purpose language such as LISP or Prolog. The OPS4 version of the language was written in LISP. OPS5, which Forgy wrote to be easier to read and maintain, has had three different interpreters, written in BLISS, MACLISP, and Franz LISP.

The OPS5 Production System

A production system is a program consisting of condition/action rules phrased in "If...then...." style. The knowledge base of an expert system written in OPS5, called "production memory," consists entirely of production rules expressing knowledge about a problem domain. OPS5 programs have two other components: a database called "working memory" and the interpreter, referred to as the "inference engine," which is the part of the system that selects and executes the appropriate rule at each point in processing.

At the start of the program, the user enters data and parameters relevant to the current instance of the problem to be solved into working memory. As processing moves along, working memory changes to reflect new information inferred by the system at each step.

The inference engine evaluates all of the rules to see which have "If" portions that are exactly satisfied by the current state of the working memory. This set of rules is called the "conflict set." If there are two or more satisfied "Ifs," the inference engine will act on whichever one its built-in protocol selects. This process is known as "conflict resolution." Examples of conflict resolution strategies are "Fire the rule with the most precise (or complex) set of conditions" and "Fire the rule that references the newest data." Upon conflict resolution, the appropriate rule "fires," that is, the "then" portion acts to change the working memory. Because this action changes the working memory, on the next round when the inference engine evaluates rules, there may be a new conflict set.

This cycle of *recognizing* the appropriate rule to fire based on the updated contents of working memory and *acting*, by firing the rule to change working memory, continues until a conclusion is reached; that is, until the conflict set is empty or a rule halts the program. The problem solution or conclusion is represented by the final state of working memory.

- *Rules Can Be Sequenced for Easy Maintenance*

The flow of control in an OPS5 program is not determined by the order in which the programmer puts the rules in the system. Rules become candidates to fire when the "If" statement is satisfied by information in working memory.

In a conventional programming language, the order of the instructions in the program is important; if you have to add a new instruction to the system, an error in its location can change the entire function to yield a wrong result.

The sequence in which rules are written in an OPS5 program is not important. Program execution does not rely on rules being in any particular order. This makes adding new information to an OPS5 program relatively easy.

- *OPS5 Works from Facts to Conclusions*

OPS5 is most often used as a forward-chaining language, which makes it appropriate for expert systems whose solutions can be reached by asking, "Given these facts, what follows?" This mode of operation has been useful in systems like XCON, the computer configuration system Professor John McDermott of Carnegie-Mellon University designed for Digital Equipment Corporation. McDermott also employed OPS5 in MUDMAN, an expert system, available from NL Baroid of Houston, for analyzing problems related to an oil-well drilling lubricant.

- *Digital's BLISS-Based OPS5 Is Fast*

OPS5 for VAX is available from Digital Equipment Corporation. This implementation of the language has a compiler and interpreter written in BLISS. BLISS is a much more efficient implementation language than LISP in terms of machine resources: a BLISS-based OPS5 executable image occupies one-tenth to one-third of the disk space that a LISP-based one occupies and has been shown to run anywhere from 4 to 30 times faster than LISP-based OPS5 programs run. OPS5 for VAX runs under the VMS operating system. It can call or be called by other VMS languages. LISP-based implementations of OPS5 designed for LISP machines and IBM-compatible personal computers are also available.

✣ Object-oriented Languages

Object-oriented programs are written in terms of "objects" rather than in terms of procedures. In this style of programming, knowledge about how to do things is associated with the objects themselves. In practice, this is done by grouping objects that do the same things in the same way into "classes" (sometimes called "flavors" or "types").

The traditional programming style is referred to as "procedure-oriented" programming. In this style of programming, a program consists of separate procedures and data. The data is manipulated by the procedures. The emphasis is on the procedures with the data serving to supply the procedures with the information they need to perform their functions.

In contrast, in object-oriented programming, the emphasis is on the data. The procedures are part of the declaration of an object (or, in practice, of a class of objects). Processing is done by requesting an object to perform a procedure in its repertoire. Two objects may perform the same procedure in different ways. For example, several objects may have procedures for adding, but an object that is a special case, one that is a range of numbers rather than a specific number, may have special ways to do addition. The procedure may, in turn, request that other objects perform other operations, or it may access part of the data "held" by the object to which it belongs.

You can do object-oriented programming in any programming language but some languages make it much easier. SMALLTALK-80,† SIMULA 67 and CLU, for instance, are designed to support object-oriented programming. In such languages, object-oriented style is the most natural way to program, and some effort must be made to use procedure-oriented programming. These languages are referred to as "object-based" programming languages.

SIMULA 67 is a language developed in the late 1960s at the Norwegian Computing Center, specifically for simulation. It has since found much more general use. Object-oriented programming was first recognized as a distinct style of programming in connection with SIMULA, and all work in this direction can be traced directly or indirectly to SIMULA.

SMALLTALK-80 and its predecessors were developed by researchers at the Xerox† Palo Alto Research Center. It is more than a language in that it incorporates a highly interactive programming environment. It was initially implemented for Xerox Corporation's own machines for internal use. In 1980, Xerox started releasing licenses and supporting code for other vendors to develop SMALLTALK-80 systems. Many such implementations were developed and SMALLTALK-80 is now available for many machines. Most of the interest in object-oriented programming was stimulated by SMALLTALK-80. SMALLTALK-80 has proven itself as an environment for developing highly interactive simulation and interactive graphics systems. There has been some promising work lately in the use of SMALLTALK-80 as a general AI development system.

Another approach to making object-oriented programming convenient is to extend an existing language. The extension is known as an object-oriented extension, and the extended language is sometimes called an object-oriented language. Usually the term "object-oriented language" is used to refer either to this type of extended conventional language or to an object-based programming language. LISP dialects are frequently extended for object-oriented programming. LOOPS, which includes an object-oriented programming facility, is an extension to INTERLISP-D. The Flavors package is an object-oriented extension to ZETALISP. The term "flavor" is simply the word used in this extension for a class of objects. COMMON LISP does not define a standard object-oriented programming facility, but most implementations are expected to provide this extension.

Some modern languages (such as Ada®) have features that, though not specifically designed for object-oriented programming, make object-oriented programming somewhat easier.

Type abstraction is very closely coupled with object-oriented programming. Separating the description of what can be done with an object (the procedures or operations that can be executed on an object) from how those procedures are implemented and, most especially, from how the object is actually represented in the computer's memory, is called "type abstraction." Type abstraction has long been recognized as a powerful and effective technique that makes it easy to modify programs during and after development. For example, a completely different implementation for a class of objects may be substituted, usually without any effect on the rest of the code. Type abstraction also makes it easier to coordinate more than one programmer working on the same program or group of programs.

"Procedural overloading" is another important feature of object-oriented programming systems. Procedural overloading refers to the fact that the same procedure (or different procedures with the same name, depending on how you want to look at it) can be implemented in different ways for different classes of objects. Procedural overloading makes it possible for a programmer to call a procedure on an object even if the programmer doesn't know to which classes the object belongs. For example, you can have a number of different classes of "graphical" objects. Each will have procedures for drawing the object, moving it, rotating it, and so on. The representation and implementation of a "circle" object may be very different from that of a "character" object, but procedural overloading allows programs like graphics editors to use the same procedures on different objects interchangeably, without the programmer worrying about how the procedure should be performed for that type of object.

Many object-based and object-oriented programming systems have a mechanism called "inheritance." In a system with inheritance one class of objects is described as a specialization of another class of objects. This has several consequences. One is that the implementation of the new class of objects can inherit part or all of the implementation of the old class of objects (its "parent" class). This can save a great deal of coding and debugging. The new parts of the new class's implementation are added only to its own definition. Because of procedural overloading, objects of the new class can frequently be used in applications that were specifically designed to use objects of the parent class without making any modification to the application's code.

Most object-oriented programming systems have been designed with prototyping in mind. The ease with which an object-oriented program can be modified and the way the final implementation of classes can be delayed make this a natural use of the object-oriented style.

Some very successful work (for instance, the CLU programming language developed at the Massachusetts Institute of Technology and an experimental programming language under development at Digital Equipment Corporation) has been done with object-oriented programming systems that have been designed for the development of production quality code. The clean organization and natural evolution from specification to high-quality code encouraged by the object-oriented programming style is seen as an advantage here.

Object-oriented programming has been shown to be very useful for graphics, animation, office applications, simulation, and modeling because this style facilitates the representation and manipulation of discrete entities. Object-oriented programming has also been very useful for knowledge representation in AI programs because object-oriented programming makes it easy to separate the knowledge from details of its representation.

Annotated Bibliography

Barr, Avron, and Edward A. Feigenbaum, (eds.), *The Handbook of Artificial Intelligence,* Vol. II, Stanford, California, HeurisTech Press, 1982. (Copyright by William Kaufman, Inc., Los Altos, California, 1981). Among other topics, the handbook discusses a variety of AI languages, including some of the early, now-obsolete languages.

Brownston, Lee, et al., *Programming in OPS5: An Introduction to Rule-Based Programming, Reading,* Addison-Wesley Publishing Company, 1985. This book is for experienced programmers who wish to develop techniques for rule-based programming in OPS5. The first part is a tutorial on OPS5 and related programming techniques. The second part compares OPS5 with other programming tools.

Campbell, J. A. (ed.), *Implementations of Prolog,* New York, (Ellis Horwood Limited, Publishers, Chichester) Halsted Press, 1984. This collection of papers presents various aspects of Prolog history and Prolog concepts and uses.

Chester, Michael, "Robotic Software Reaches Out for Task-Oriented Languages," *Electronic Design,* May 12, 1983, pp. 119-129. This article discusses languages and software packages that offer some advances in the effort to develop robot programming tools.

Clark, K.L., and S.A. Tarnlund (eds.), *Logic Programming,* New York or London, Academic Press, 1982. This book comprises twenty-three papers on subjects relating to logic programming in general and to Prolog in particular.

Clocksin, William F., and Jon D. Young, "Introduction to Prolog, A 'Fifth-Generation' Language," *Computer World,* August 1, 1983, pp. 1-16. This article presents a clear, brief description of the Prolog language with simple examples of coding and a discussion of applications.

Evanczuk, Stephen, and Tom Manuel, "Practical Systems Use Natural Languages and Store Human Expertise," *Electronics,* December 1, 1983, pp. 139-145. In the context of practical applications, the authors briefly discuss PROLOG and SMALLTALK, LISP, and AI program development environments. Some discussion of expert systems and natural language systems is included.

Forgy, Charles L., *OPS5 User's Manual,* Department of Computer Science, Carnegie-Mellon University, Pittsburgh, 1981, Charles L. Forgy. This is an introduction to, and reference manual for, the OPS5 production system programming language.

Fox, Jeff, "Big Ideas from SMALLTALK," *PC World,* March 1984, pp. 72-75. This article discusses object-oriented programming and technologies such as multitasking, concurrency, windows, and mouse interface devices in the context of the Smalltalk language.

Freedman, Roy S., "The Common Sense of Object-Oriented Languages," *Computer Design,* February 1983, pp. 111-118. This article tells what object-oriented programming is and discusses its utility.

Friedman, Daniel P., *The Little LISPer,* 1974, Science Research Associates, Inc., Chicago. This is an enjoyable introduction to LISP programming.

Gevarter, William B., "The Languages and Computers of Artificial Intelligence," *Computers in Mechanical Engineering,* November 1983, pp. 33-38. This article presents a discussion of LISP, Prolog, and other AI programming languages and associated hardware.

Kowalski, J.A., "Logic Programming—Past, Present and Future," *New Generation Computing,* Vol. 1, 1983, pp. 107-124. This article provides a historical and conceptual overview and projections for the future of logic programming, including Prolog.

Palmer, Richard, "LISP and Artificial Intelligence," *ICP Interface,* spring 1983, pp. 16-22. This article puts LISP into a historical context and describes how it will affect the electronic data processing business over the next 10 to 15 years. The article includes examples of LISP functions.

Pugh, John R., "Actors Set the Stage for Software Advances," *Computer Designs,* September 1984, pp. 185-189. This article discusses the actor (object) oriented paradigm for programming in terms of computer architecture, data abstraction, the behavior of systems built on this paradigm, and applications suited to it.

Rauch-Hindin, Wendy, "Speak the Language," *Systems and Software,* December 1983, pp. 174-177. This article offers a comparison of LISP and Prolog and a discussion of Prolog features and uses.

Rich, Elaine, *Artificial Intelligence,* New York, McGraw-Hill Book Company, 1983. Among other topics, Rich's book discusses the desirable features in AI languages in general and describes particular AI languages, past and present.

Robson, David, "Object-Oriented Software Systems," *Byte,* August 1981, pp. 74-86. This is a clear, easy-to-understand explanation of the basics of object-oriented programming. Special note: This issue of *Byte* was devoted to object-oriented programming. It included a discussion of SMALLTALK.

Steele, Guy L., Jr., *COMMON LISP: The Language,* Maynard, Massachusetts, Digital Press, 1984. This is the standard reference for COMMON LISP. The book contains many examples of the use of COMMON LISP functions. Implementation notes suggest techniques for handling tricky cases. Compatibility notes compare or contrast COMMON LISP features with those of other widely used LISP dialects.

Symbolics 3600 Technical Summary, May 1984, Symbolics, Inc., Cambridge, Massachusetts. This document describes the Symbolics, Inc., LISP machine. The Flavors features are discussed on pp. 49-61.

Winston, Patrick H., and Berthold Klaus Paul Horn, LISP (2nd ed.), Reading, Massachusetts, Addison-Wesley Publishing Company, 1984. This book gives the basics of LISP programming and demonstrates how LISP is used in actual practice. This second edition features COMMON LISP, the new standard for the language.

Notes

1. Rich, Elaine, *Artificial Intelligence,* New York, McGraw-Hill Book Company, 1983, pp. 392-401.

2. Chester, Michael, "Robotic Software Reaches Out for Task-Oriented Languages," *Electronic Design,* May 12, 1983, pp. 121-122, 124.

Chapter 5

**Computer Hardware
for Artificial Intelligence**

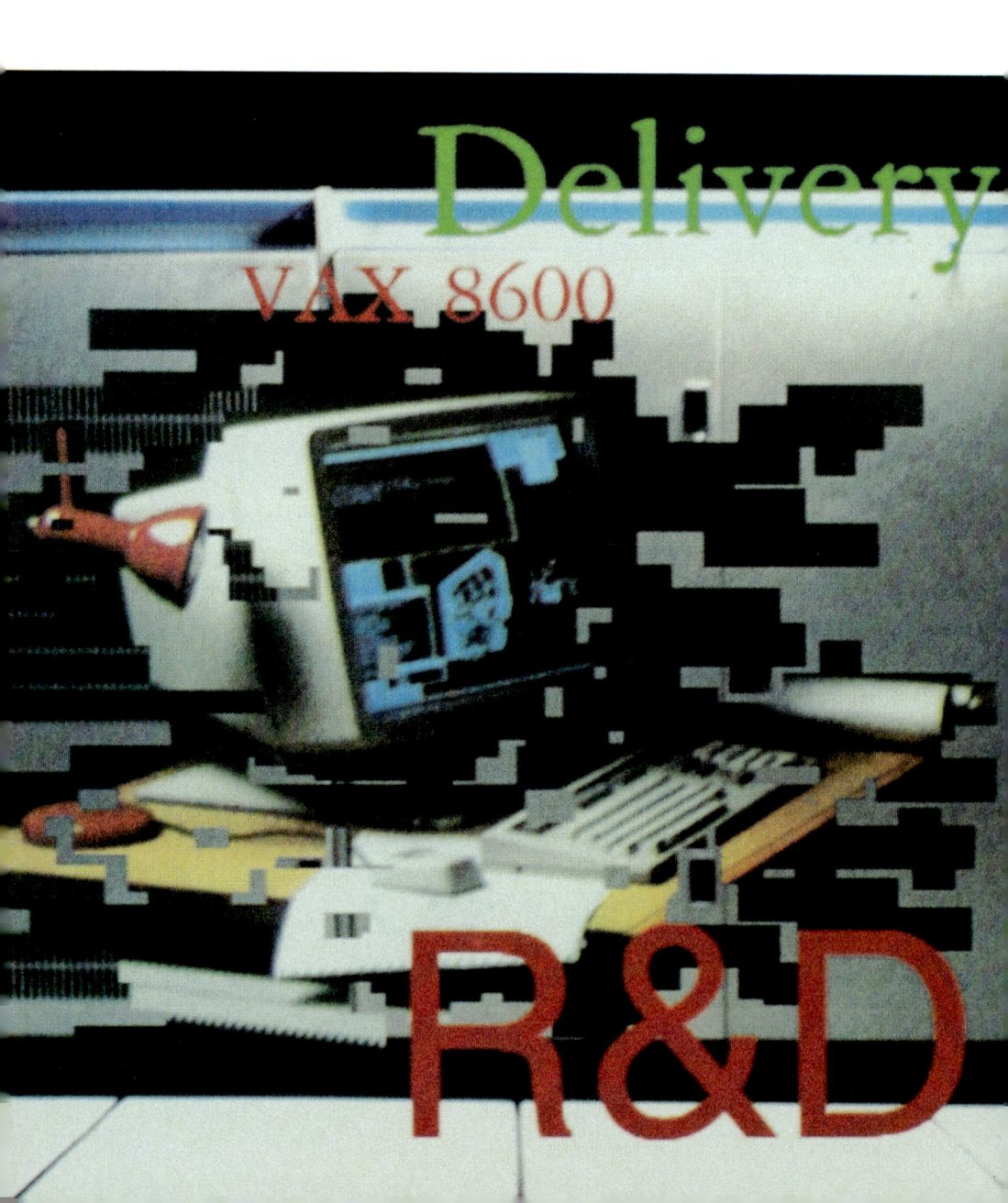

Computers used in artificial intelligence work can be broken into three main categories:

- Those used for the development of practical applications.
- Those used for delivering practical applications to end users.
- Those used for conducting research and development.

The criteria for buying hardware to support artificial intelligence work differ for the three categories. Performance is generally the most important criterion for computers used in research and development. This is because the rate of progress in research work is strongly influenced by the ease with which researchers can interact with the computer and by the rate at which they can obtain feedback from the developing system. Price is usually considered the more important factor in the purchase of delivery hardware because at the delivery phase, features that support program development are not necessary but the ability to run the application economically in multiple locations often is. As in research and development, the application development process is best served by hardware with good performance characteristics and good programming facilities. But the choice in hardware is often dictated by which hardware will eventually be used for delivering the application, a decision influenced by cost considerations.

Factors That Apply to All Categories

Artificial intelligence programs frequently require a large address space. A large address space is an advantage because it allows the program to address many locations without requiring the hardware or software to do a lot of context-switching, which would slow down processing. Processing is simpler, cleaner, and faster if you have a large address space.

A word size of 32 bits is likely to be adequate for the near future. Thirty-two bits should allow you to address all the objects you might want to use in any sizable program. A smaller word size is often insufficient.

Today's popular personal computers have less than 1 Mbyte of physical memory, which is not enough for most expert systems. However, personal computers with larger memories of up to 3 Mbytes are being introduced, and they should be capable of supporting "real," albeit small, AI systems. Five to 20 Mbytes of physical memory ought to be enough for most systems, but as one of Digital's AI developers has noted, "You learn quickly never to limit yourself—there are always bigger things you want to do."

Artificial intelligence programs are characterized as consuming a lot of machine cycles compared to traditional programs. This is primarily because AI programs tend to tackle complex problems that require operations like searching or pattern-matching. For this reason, speed is likely to be more of a consideration in a computer you use for AI work than for one you use in regular data processing.

As in traditional computing, software availability is an important consideration in your AI hardware selection. You may want to be able to utilize knowledge-engineering software and many traditional languages and software packages in addition to artificial intelligence languages within AI applications. It is also important to choose hardware that runs the particular dialects of the AI languages that suit your development style.

If your organization is working on a variety of problems on a variety of computers, networking will be desirable in your choice of hardware for both development and delivery. With good networking, developers at different sites can share code and files efficiently and distribute applications and updates in a timely fashion. Networking can also be important from the standpoint of delivery because it allows you to distribute the application quickly and conveniently and access centrally stored data.

Systems for Practical Application Development

Most industry problems that involve any artificial intelligence are mixtures of problems for which AI solutions are appropriate and problems for which traditional solutions are appropriate. Applications have already been developed to solve many of the common problems that traditional programs address using general purpose hardware already in place throughout the business world. It is not necessary to throw away the solutions your organization has already spent considerable time and money achieving. If just a piece of the solution requires an artificial intelligence approach, perhaps something written in LISP, a general purpose computer with a common language interface lets you utilize the traditional language solution along with the new LISP code; they can work together as equals. A general purpose machine also provides database management facilities and I/O throughputs.

An important consideration in choosing hardware for application development is the ease with which you can move from development to delivery. Compatibility between systems or using the same system for development and delivery ensures that the application can be run without modification.

Of course, the usual criteria for hardware purchases apply — price; service; reliability; ease of installation and maintenance; the capacity to upgrade the system; and vendor support, training, and consulting.

General Purpose Computers for Application Development General purpose computers may be the best hardware investment when

- In addition to doing research and development, you want to deliver AI programs to real world applications.
  You wish to integrate AI code with traditional algorithms in the same program.
- Your application requires multiuser access or concurrent access to a large database.
- Your organization also intends to create types of applications other than AI programs.
- A supercomputer is either more hardware than is needed or than is affordable.
- You already have a general purpose computer and therefore prefer to utilize it instead of making an additional hardware purchase.

If your organization does traditional computing on a general purpose minicomputer, mainframe, or workstation, you may not need to invest in a specialized artificial intelligence workstation. Most AI languages and development tools have been implemented for use on general purpose computers, so the general purpose computer gives you the flexibility to do both traditional computing and artificial intelligence on the same machine. This flexibility also means that if you want to learn to use artificial intelligence languages and techniques, you don't have to invest in special hardware.

With a general purpose computer, you can mix artificial intelligence techniques and conventional tools efficiently in a problem solution. You're not locked into AI techniques or a LISP-optimized architecture. Moreover, a wealth of traditional software for developing those non-AI algorithms is available on general purpose computers.

Characteristics to look for in a general purpose machine are

- Program development tools like language-sensitive editors, code management systems, and multilanguage symbolic debugging facilities.
- A 32-bit virtual address space.
- Screen-oriented utilities, the capability to manipulate multiple windows, and other programmer productivity enhancements.
- A good mix of standard development languages with proven track records and consistency in languages so that they can all use a common language interface and existing operating system services and facilities.
- Database systems accessible from all languages. The ability to mix a variety of database management options including standard CODASYL and relational database systems that you can use in any mix.
- Demonstrated large networks, including the ability to communicate with other vendors' machines, so that people can easily share in the incremental development work and can eventually distribute delivery.

Mainframes Most of the research and development in artificial intelligence, until recently, has been done on mainframes. In fact, the first computer designed specifically for LISP execution was the PDP-6 (the predecessor of the DECSYSTEM-10 and DECSYSTEM-20) a mainframe introduced by Digital Equipment Corporation in 1964.

Mainframes offer speed and flexibility for doing traditional computing, as well as artificial intelligence. Common language interfaces also let you mix traditional computing and artificial intelligence in the same system.

Recently, the price to performance relationship of minicomputers and workstations have encouraged many artificial intelligence developers to move most interactive work to smaller computers, reserving the mainframe for batch computing and database support related to the development process.

Minicomputers Minicomputers offer speed, a large address space, and flexibility at a moderate cost per user. When you're doing both traditional computing and AI work, a minicomputer, like a larger general purpose system, lets you mix the two, but at a lower price.

The wide range of traditional software available to run on these machines is valuable in algorithmic portions of artificial intelligence solutions, and a variety of AI languages has now been implemented to run under different operating systems.

Many artificial intelligence workers believe they need a dedicated computer system in order to get the computer resources they require. The low price of a minicomputer can make this possible.

General Purpose Workstations A growing trend for AI development (and research, as well) is a general purpose workstation.

The advantages of a workstation are

- It can use all the computing power and facilities needed for optimal performance on a single AI task.
- All the computing power and facilities are available for the AI task and for a single user. This assures predictable, optimal performance.
- An individual workstation assures sufficient resources to support bit-mapped graphics, which require a large amount of processing power.
- Many general tools and utilities are available—MAIL, UNIX systems, and UNIX-like utilities, a choice of editors, standard text-processing packages, standard graphics libraries, and floating point performance.
- Like minicomputers and mainframes, a general purpose workstation is good for delivery as well as development.

An artificial intelligence workstation should have a high-resolution screen; windowing with the ability to create, fill, move, reshape, and overlap; a convenient interface tool such as a mouse; a good editor; a good debugger; and tools designed for programming the system in the language of your choice.

- *Systems That Are Useful for Delivering Practical Applications to End Users*

   Delivery systems include primarily mainframes, minicomputers, general purpose workstations, and personal computers. The considerations for choosing delivery systems are much different from those for research and development.

   While speed is important in applications that must operate in time-critical environments, a more important consideration is cost per user. Especially if you have an expert system you intend to make widely available, you will want an inexpensive delivery vehicle. Traditionally, applications have been delivered on timesharing systems. However, the low cost and the user-friendly interfaces of some of the personal computers make them good delivery vehicles for some small AI applications.

   Although workstations are generally thought of as research and development machines, some of them on the less expensive end of the scale are appropriate as delivery vehicles. If you are already doing development on a general purpose workstation, you may want to use the same operating system for delivery in order to ensure compatibility. In fact, the ideal is to do the development and the delivery on the same kind of system.

   Most delivery systems require interactive capabilities. This means responsive systems with large screens and, depending on the application, graphics.

   Networking capability is an especially important consideration at the delivery phase, so that the application can be efficiently distributed and so that expensive resources can be shared.

- *Computers for Research and Development*

   Research and development (R&D) computers span a wide range from supercomputers down through specialty machines (LISP machines), general purpose mainframes, minicomputers, and workstations.

   There are several cost, performance, and software factors to consider in choosing among these different types of computers for AI research and development purposes. However, the primary factor in purchase decisions for hardware intended for R&D in artificial intelligence is performance.

As mentioned above, artificial intelligence programs generally require more machine cycles than traditional programs do. Because the time required for the research and development phase of a project depends partly on how fast you get feedback, speed is an important consideration in choosing an R&D system.

The normal requirement for a large physical memory is even more pronounced during the research and development phase of artificial intelligence programs because at this time the system requires the use of a debugger, a compiler, and an editor. These facilities require large amounts of disk space, virtual memory space, and/or many machine cycles. However, the actual amount of memory required depends on the application.

All artificial intelligence development work (and many delivery systems) require a good deal of interactive capability. This translates to responsive systems with large, high-resolution screens, a predictable response time, and an adequate I/O bandwidth.

If you want to develop artificial intelligence applications with input from several sources, and if you want to be able to easily distribute the incrementally developed results for testing and further refinement, it is important to do the work on computers that are supported by strong network capabilities. Networking allows the developer to start processes on several systems and tap remote files or programs. Dialup and other telecommunications features also allow developers to work from home and remote sites, as well as at the office.

And finally, the languages available to run on your R&D hardware are very important. It is an advantage to use a system that can efficiently run traditional software that interfaces compatibly with the artificial intelligence software you will be using. This compatibility allows you to make use of traditional methods to develop algorithmic portions of AI programs and use already existing programs and databases within a larger AI program.

In addition to considering the range of languages available to run on your R&D hardware, it is important to choose hardware that runs the particular implementations of languages that suit your purposes. For instance, in recent years, some developers have preferred to work in MACLISP, which is considered faster but does not have as many built-in programmer support facilities as an alternative preferred by some other developers, INTERLISP, which is considered to be much slower.

Supercomputers for R&D While high-speed numerical processors like the CRAY-1® and the CDC CYBER® series have the size, power, and speed to run AI programs fast, present-day "supercomputers" generally are used for AI only in very specific problems that require their great speed. This is because supercomputers don't have the types of operating systems and software that are conducive to AI development, nor are they widely available.

Workstations Specialized for R&D: LISP Machines Workstations designed to perform symbolic processing and optimized for running the LISP language have been on the market since 1981. LISP machines are dedicated workstations with hardware architectures designed to run LISP. These are powerful, single-user machines with large physical memories, generally 4 to 16 Mbytes. LISP machine configurations generally cost in the range of $80,000 to $170,000.

The advantages of a computer that is optimized for LISP stem from its large physical and virtual memory and method of memory management, from its speed, and, most importantly, from its having an architecture optimized for running LISP. Current LISP machines also have bit-mapped, high-resolution graphics and overlapping windows, a feature that is particularly helpful for program developers. Windows allow the system to represent the discrete activity of more than one process on a single screen as if you had more than one terminal at your desk. Multiple windows allow programmers to switch from one environment to another without losing their contexts. Graphics are useful in that they are frequently used in AI development to illustrate information and relationships.

The different LISP machines incorporate various AI programming tools, such as LOOPS and Flavors packages. These tools, coupled with LISP execution features designed into the hardware, are often significant advantages in developing large programs and for developing prototypes quickly.

In evaluating a LISP machine, note the LISP dialect from which it was implemented—ZETALISP or INTERLISP—the development tools it can run, and the availability of other programming languages for it, as well as speed.

The first LISP machines were built at the Laboratory for Computer Science at MIT and at the Xerox Corporation Palo Alto Research Center. The Xerox 1100,† based on INTERLISP, was the first LISP machine on the market, in 1981. Symbolics, Inc., and LISP Machine, Inc., developed their machines from the MIT LISP machine, and are among LISP machines currently available on the market.

The Case for Basing Your AI on Digital Equipment Corporation Systems

General purpose computers are not always the perfect choice, as you have seen, but more often than not they are the best choice. The strength of general purpose computers is precisely their general purpose nature—their flexibility. Not only do they provide broad support for program development, rather than specialized support for a particular language, but they also have operating systems that are more mature with a richer offering of user services, more sophisticated and flexible networking capabilities and, most importantly, a realistic opportunity to deliver practical artificial intelligence applications.

No system embodies the virtues of general purpose computing for artificial intelligence better than VAX systems. Digital's VAX superminicomputers are unique, allowing you to do research and development, application development, and delivery all on one system. When you want to do practical AI work, you need to consider all three of these functions.

As you recall, large amounts of memory are critical to most artificial intelligence applications and, given today's technology, virtual memory is the only practical way of getting it. "VAX" stands for Virtual Address extension. It was designed from the ground up to run virtual memory systems. With the VMS (Virtual Memory System) operating system, processes can address *billions,* not just millions, of memory locations—certainly sufficient for any AI application you're likely to be working on.

The VAX architecture has been implemented in computers ranging from a high-performance microprocessor-based workstation, the VAXstation II, to a mainframe-level system, the VAX 8600 (with over 4.5 million instructions per second). If that's not enough performance, you can cluster as many as 16 VAX processors in a loosely coupled multiprocessing VAXcluster configuration. VAX 8600 is the only multiuser system that executes LISP at LISP machine speeds or faster. With VAX superminicomputers, a single *compatible* line of computers can meet practically any demand for research, development, and delivery.

No system is better equipped for program development than the VAX system. Digital offers two operating systems—the VMS operating system, widely acclaimed for its sophisticated features for efficient program development; and the ULTRIX operating system, Digital's extended implementation of the UNIX† operating system that has been particularly popular among researchers. (the UNIX operating system, by the way, was originally developed at AT&T Bell Laboratories on Digital's PDP-7 computer.) Between these two operating systems, practically every major programming language is supported, including FORTRAN, COBOL, BASIC, PL/1, PASCAL, Ada® and, of course, VAX LISP, INTERLISP, Quintus Prolog, Prolog II (available in Europe), OPS5, and the knowledge-engineering tool ART.†

Other AI languages, dialects, and tools are available from third parties. VMS systems also include a wealth of information management tools like VAX DBMS, a database management system; VAX Rdb/VMS, a relational database management system; DATATRIEVE for rapid, flexible data retrieval; the FMS forms management system; VAX TDMS, a terminal data management system; and the VAX Common Data Dictionary. All of these facilities, and the powerful VMS system services and record management services, are callable from any of the Digital-developed languages. In fact, the VMS common language environment assures that routines written in any VMS language are callable from any other language. Add to all of this the optional code management software that is indispensable when you have large development projects involving many programmers and large numbers of modules. You can see why VAX systems are widely considered the premier development systems.

Digital's VAX-based workstations add the features so critical to AI research and development—sophisticated, high-resolution graphics; multiwindowing and multitasking; and an easy and efficient human interface. The VAXstation II is a fully integrated workstation that runs on the new MicroVAX II microprocessor; it puts VAX-11/780 class performance at a developer's desk with 1 to 9 Mbytes of memory, dedicated disk, and high-resolution graphics.

While embedded development tools may enhance speed somewhat for R&D purposes, Digital's standard graphics languages, GKS and DECOR provide excellent graphics development features. You can call them from any Digital language—including LISP. Using industry-standard graphics languages provides another benefit—device independence. Graphics you develop with GKS or DECOR on one graphics subsystem can be run on another subsystem with little or no modification, similar to the high-level programming languages themselves. Digital's graphics standards also make graphics systems accessible to all parts of a large application written in several computer languages.

When it comes to networking, again, VAX systems are without peer. VAX systems offer every conceivable networking topology, from baseband, broadband, and fiber optic local area networks to global satellite networks. VAX users and programs, using Digital's special DECnet networking system, can communicate with DECSYSTEM-20, PDP-11, and Rainbow and IBM personal computers with remarkable ease. DECnet also works under the ULTRIX-32 operating system. For the best in computer networking, there's VAX-to-VAX. It's so good, you would hardly know it's there. VAX systems can also communicate with most other manufacturers' systems through such communication devices as SNA and X.25 gateways, to name just two.

Although VAX superminicomputers are the best systems for most AI development and delivery, Digital offers other systems that might serve your needs.

The DECSYSTEM-20 has proven itself in AI research for 20 years. From the mid-1960s on, AI researchers have made use of the large virtual address space and the processing power of the DECsystem-10 and later, the DECSYSTEM-20. The TOPS-20 operating system for the DECSYSTEM-20 provides a comfortable user interface as well as facilities useful in AI program development. The DECSYSTEM-20 can run INTERLISP, MACLISP, Portable Standard LISP, and DECSYSTEM-20 Common LISP (CLISP).

Digital's Rainbow personal computer also runs artificial intelligence languages: Golden Common LISP (GCLISP) and Prolog II (available in Europe). When you are first getting started with AI, a personal computer like Digital's Rainbow provides an opportunity to try out AI techniques inexpensively. For instance, if you want to try writing and running some LISP programs, GCLISP runs on Digital's Rainbow personal computer and on other compatibles. GCLISP, at a list price of under $500, comes with the San Marco LISP Explorer,† a LISP tutorial written by Professor Patrick H. Winston. Thus, the Rainbow offers an inexpensive way for programmers to learn basic AI techniques and to gain experience with simple AI programs.

The programs you develop on the Rainbow in GCLISP are source compatible with VAX LISP programs. Because they can be transported to more powerful hardware systems, programs developed on the Rainbow can be used for real AI, in addition to training.

❋ Annotated Bibliography

Deering, Michael F., "Hardware and Software Architectures for Efficient AI," *Proceedings of the National Conference on Artificial Intelligence, 1984,* University of Texas at Austin,© 1983, The American Association for Artificial Intelligence, distributed by William Kaufman, Inc, Los Altos, California. This paper discusses efforts to improve AI computational throughput and reduce costs through improvements in hardware and software.

Fahlman, Scott, "Computing Facilities for AI: A Survey of Present and Near-Future Options," *AI Magazine,* winter, 1980-1981, pp. 16-23. Though Fahlman noted that this article would be out of date by 1982, it still provides a good discussion of important considerations in choosing computing facilities for AI.

Manuel, Tom, "Lisp and Prolog Machines Are Proliferating," *Electronics,* November 3, 1983, pp. 132-137. This is a survey of AI hardware systems and research. The article discusses the origins and features of LISP machines and briefly discusses features of general purpose computers and workstations.

Mokhoff, Nicolas, "Parallelism Makes Strong Bid for Next Generation Computers," *Computer Design,* September 1984. This article discusses how parallel architectures will affect supercomputers and fifth-generation, knowledge-based computers.

Rauch-Hindin, Wendy, "Artificial Intelligence Coming of Age," *Systems and Software*, August 1984, pp. 108-118. This article discusses specialized hardware and software for developing AI applications, with a special focus on the Symbolics LISP Machines.

Spitznogle, Frank, "Practical Tools Earn a New Level of Respectability," *Computer Design*, September 1984, pp. 197-200. This article discusses advanced architectures that include hardware and software systems.

"Targeting AI Hardware: How Machines Mix in the Marketplace," *Target; The Artificial Intelligence Business Newsletter*, Vol. 1, Issue 1, March 1985, pp. 1-6. This article discusses the artificial intelligence target markets that current computer vendors are reaching and the directions in which vendors are developing their products.

Chapter 6

**Getting Started with Expert Systems:
A Case Study**

Knowledge engineering

As we saw in Chapter 2, AI applications are up and running in a variety of areas. The first practical applications were developed by universities or other research institutions as custom designs for, or as joint research projects with, a corporation or an agency of a national government. With the demonstrated success of these applications, many companies are creating commercial products, both for internal use and for sale.

At this time, expert systems is the area of AI that offers the most immediate prospect of economic rewards. Now is a good time to investigate expert systems solutions because, although expert systems require a lot of computing power, the cost of this resource is constantly decreasing, while human expertise remains expensive. Moreover, languages, hardware, and specialized tools for developing expert systems are now commercially available.

If you're asking "What can an expert systems approach do for a problem my organization is confronting?" and "What would it take to develop an expert system?" this chapter will provide you with information and opinions from people who have put expert systems to work in a commercial environment. First, we will look at the experience of Digital Equipment Corporation in the development of the expert system XCON. We will examine potential pitfalls, rewards, and recommendations that have been cited by Digital employees involved in the management, development, and integration of XCON. Then we will review the opinions of these people on the types of problems expert systems are good at solving and what the development process entails. And finally, we will list sources of assistance and training in AI and expert systems technology.

It should be noted that this chapter describes the expert system development process in the light of Digital's experience. It is not a prescription for building all expert systems. Nevertheless, you may find Digital's experience instructive and thought-provoking.

Digital Equipment Corporation and Expert Systems

Digital Equipment Corporation uses several of the world's largest expert systems on a regular basis and is in the process of developing many more. All of these systems to date have been built using VAX-11/780 and -750 systems under the VMS operating system. The first to be developed, and the largest commercial expert system in the world used on a daily basis, is XCON, for eXpert CONfigurer of computer systems. (XCON is also known as R1 in the research community.) A configurer generates a configuration, a detailed plan for how a selection of computer components should fit together.

The Business Problem

Today, most Digital Equipment Corporation computer systems are custom tailored VAX superminicomputer systems, commonly utilizing from 10 to 200 orderable parts from among 8,476 items of hardware, software, accessories, documentation, and services recognized in the database used by XCON (as of May 1985). Further complicating the configuration task, VAX systems are configured not only individually but also on cluster, network, and multiprocessor levels.

The configuration components include central processing units, memory, tapes, disks, adapters, cables, terminators, bus repeaters, processor options, cabinets, boxes, power supplies, regulators, terminals, software, and many select groupings of each that are offered via a single Digital option designation, all of which come in many different variations.

Because a large proportion of VAX systems are unique configurations, there is no standard configuration for the system components. Many system combinations constitute a new problem, one among possible combinations numbering in the millions, even if the choices are restricted to the common options only. And that makes the configuration process itself a problem.

Each system's components must be fitted into a detailed plan. Fitting them into the plan is a complex design task. To create this plan, the configurer must have all the basic system engineering data and all the rules for the configuration components at hand and be able to apply all of these when doing the detailed plan.

The need for a configuration system to manage these complexities became evident in the mid-1970s in configuring PDP-11 computer customer orders. It was foreseen that increasing volume could strain Digital's capacity to perform this immense configuration task within reasonable constraints on the manufacturing organization. Digital management at that time knew that VAX superminicomputers would be Digital's main revenue generators by the 1980s. Therefore, they decided initially to focus the configuration system on VAX systems instead of on PDP-11 computers. VAX computers were also chosen as the focus of XCON because the number of peripheral components that could be configured with a VAX at its time of introduction was not so large as to make the project unmanageable. In addition, the experts who would need to supply the configuration information to the developers were close to each other, making knowledge acquisition easier.

- *Why Didn't Digital Use Conventional Methods to Solve the Problem?*
In 1974, a design engineer in Digital's PDP-11 Systems Engineering organization suggested that an automated technical editor should be developed to check orders. Some work was done toward implementing such a system. In early 1976, Dennis O'Connor, now Senior Group Manager for Intelligent Systems Technologies Group, began his involvement with XCON. As Group Engineering Manager of Manufacturing Engineering, O'Connor provided Richard Caruso, an engineer in PDP-11 Systems Engineering, with the support to conduct a thorough analysis of the configuration problem and its impact on various organizations. Caruso suggested that the initial effort to develop an automated checker be developed into a full-blown automated configuration system as a means of setting up a configuration discipline. Before attempting an AI solution to the configuration problem, two subtasks were automated as traditional systems, one in FORTRAN and the other in BASIC. Following approval of Caruso's problem definition, a lengthy specification process followed, but the automated subtasks were never tied into a completed traditional system.

Some people feel that a traditional program could have been developed that would perform the configuration task. Others feel that a difficult obstacle for any traditional system would have been the fact that a solution had to aim at a moving target. New components were constantly being added, and knowledge about the components was growing and changing, so the program would have been subjected to a constant need for revision. Traditional programs are not flexible or easily modifiable, while AI programs are more amenable to ongoing development, so the need for revision would have meant a major overhaul for a traditional program. One engineer who has worked on the problem from the beginning gives credit for the success of XCON both to Professor John McDermott, who developed the prototype, and to very practical AI techniques, which McDermott utilized. Most of Digital's development team agree that the AI solution has been proven and is ultimately the best approach for this problem.

- *AI Methods Worked*

The engineers who had performed the configuration task manually could put a certain amount of their knowledge into writing. However, much of their knowledge consisted of rules of thumb that they could apply on the spot, but that they couldn't always articulate when trying to define configuration procedures. Even if they had been able to articulate all of the rules, this type of knowledge does not lend itself to programming in a sequential language. But an AI language offered a solution in the form of production rules. Production rules, used in many expert systems to express "situation/action" knowledge, accommodate configuration rules of thumb easily.

Another reason an AI solution worked well was that an expert system was able to search through the problem by means of heuristics without knowing all the paths exactly. In general, when you don't know all the paths and how to express the knowledge, AI and expert systems allow you to capture knowledge as it becomes available and put it to immediate use.

The AI techniques used in XCON also made it easy to develop definite general solutions to the configuration subtasks while allowing for the straightforward addition of more specific solutions for exceptional cases. Configuration is, for the most part, a complicated pattern matching problem. The OPS (OPS5, discussed in Chapter 4, was preceded by OPS4) language contains a very powerful inference engine that freed the programmers to pursue the real problems of the configuration task. This allowed the developers to quickly build a core system that they could extend as exceptions were added. It was easy to add new knowledge to an expert system as the engineers articulated more and more knowledge. This ease resulted from the fact that the OPS4 (and later OPS5) language in which XCON was written did not require the code for production rules to follow any particular sequence, so the programmers could insert new code where it was most convenient for them.

An AI solution did not come totally out of the blue for Digital. Most AI research in the 1970s had been done on Digital's DECsystem-10 and DECSYSTEM-20 computers, and Digital had exchanged both engineers and information with academic AI research laboratories. This experience created a more receptive atmosphere than might otherwise have been the case when an artificial intelligence approach was suggested. Some people were skeptical that an AI solution could work. But many other people believed in, were committed to, and supported the project throughout its development. XCON's success rewarded its supporters and changed the skeptics into believers. Now dozens of additional AI projects are in the works at Digital.

- *The Development Process*

In 1978, a Carnegie-Mellon University Computer Science Department professor who came to work for Digital suggested that a knowledge-based system might solve the problem of automating the configuration task. This led to Digital's agreement to support a CMU project to develop XCON. CMU's John McDermott began working with his colleagues on XCON in December 1978. (XCON began to be known in the research community as R1 in August 1980.) The original task was to configure VAX-11/780 computer systems. In approximately a year, XCON developed from an idea to a basic prototype to a system with true expertise.

The system went through the following development phases:

- Strategy formulation, initial prototype construction, and demonstration—December 1978 through April 1979. During this period, McDermott learned the basics of VAX configuration and implemented a prototype version, equipped with about 250 rules. Richard Caruso delineated the structure of the task and described the actions each subtask needed to perform.

- Large scale knowledge acquisition—May 1979 through September 1979. McDermott consulted with Caruso to fill in gaps in the prototype's knowledge and to refine the knowledge, eliciting information on exceptions that had not been articulated earlier.
- Testing—October through November 1979. XCON was given 50 orders to configure, and a team of configuration experts from manufacturing and engineering evaluated the output for errors. With corrections to some rules and correction of an architectural problem, all the orders came out right. XCON was judged to function well enough to perform the configuration task. At this point, the system had about 750 rules.
- During this period, planning was also done in order to integrate the system into the existing organizational structure. Processes were put in place for adding new product descriptions to the database and to add to or modify rules.
- First installation—January 1980. XCON was installed in a manufacturing plant in Salem, New Hampshire, and began to be used on a daily basis, closely supervised by the technician who had previously been doing the task.
- Integration and utilization of the system—June through December 1980. XCON began to be used in more and more of Digital's final assembly and test plants. Digital software engineers extended the system to configure VAX-11/750s at this time, also.
- Reimplementation of XCON in OPS5—completed in July 1980. Dr. Charles Forgy of Carnegie-Mellon University, the principal designer of the OPS4 language in which XCON was originally written, developed OPS5. The new version of the language was easier for developers to read, and hence easier to maintain; it ran faster; and it was VAX-based, which was a strategic decision on the part of Digital. When McDermott and his colleagues at CMU reimplemented XCON in OPS5, some redundancies were also eliminated and the program was, in general, streamlined with the transfer of LISP algorithms for some subtasks to OPS5 rules. The reimplementation allowed Digital to convert XCON from running on a DECsystem-10 computer, which OPS4 had required, to running on a VAX system.
- Utilization of the system and ongoing revision of the knowledge base under Digital management—January 1981 to July 1981. During this time, the system was extended to configure VAX-11/730 and PDP-11/23 computers. XCON's capabilities were refined to improve UNIBUS con-

figuration, to perform more cabling, and to compute addresses and vectors for peripheral devices.
- Conversion of XCON to a BLISS-based version of OPS5 – January 1982. This conversion was done entirely by Digital personnel and was the final transition of XCON from CMU to Digital. By this time, Digital had become self-sufficient in continuing the development of XCON, and CMU had settled into a consulting role.
- XCON was extended to configure all Digital systems sold in significant volume – late 1982 through November 1983. In addition, a production-mode outer shell was added. The shell contained embedded statistical and performance monitoring capability, a user interface layer, a connection to a database management system order database, and a problem-reporting subsystem. In early 1985, the system had about 4,200 rules.
- Creation of a formal inhouse training program – January 1983. This 14-week training program was formed to develop proficiency in OPS5 and LISP, to provide familiarity with AI techniques and tools and knowledge engineering, and to train AI engineers to design, develop, and implement inhouse systems, including XCON. The program now also trains Digital employees to manage AI development programs.

- *How XCON Works*

XCON accepts as input a list of items on a customer order, configures them into a system, notes any additions, deletions, or changes needed in the order to make the system complete and functional, and prints out a set of detailed diagrams showing the spatial relationships among the components as they should be assembled in the factory.

The users are

- Technical editors who are responsible for seeing that only configurable orders are committed to the manufacturing flow.
- The assemblers and technicians in Digital's manufacturing organization who assemble the systems on the plant floor.
- Sales people, who use XCON in conjunction with XSEL, an expert system that helps them prepare accurate quotes for customers. This can be done on a dialup basis from the customer's site.
- Scheduling personnel who use information from XCON to decide how to combine options for the most efficient configurations.
- Technicians who assemble systems at the customer's site.

Configuration tasks like that of XCON can be thought of as heuristic searches for an acceptable configuration through a search space of possible configurations. McDermott used a "match search" pattern-matching method that does not deviate from the solution path; backtracking is rarely ever required because XCON's knowledge is usually sufficient to determine an acceptable next step.

Elements of the system include

- OPS5's production memory (knowledge in condition/action rule form about how to configure the systems), the embodiment of the heuristic knowledge base.
- The working memory, which starts with the set of customer-ordered components but by the end of processing has accumulated descriptions of the components and descriptions of partial configurations that will be used to complete the full configuration.
- The inference engine, which is the OPS5 interpreter. The interpreter selects and applies rules.
- An additional component database (descriptions of each of the components that may be configured in systems).
- User interface software that allows the user to interactively enter and modify orders, review XCON output for those orders, and enter problem reports.
- Traditional software for database access, the collection of statistics on hardware resource utilization and functional accuracy, and for automatically routing problem reports entered by users to the support organization.

The major subtasks within XCON are

- Checking the order for gross errors, such as missing prerequisites, wrong voltage or frequency, no central processing unit, etc. The first subtask is also concerned with unbundling line items to the configurable level, assigning devices to controllers, and distributing modules among multiple secondary buses.
- Placing the components in the central processing unit cabinets and then finding an acceptable configuration of the secondary bus by placing modules in backplanes, backplanes in boxes, and boxes in cabinets.

- Configuring the rest of the components on the secondary bus — panels, continuity cards, and unused backplanes. Also, computing vector and address locations for all modules.
- Laying out the system on the floor and determining how to cable it together.

The production rules in XCON that describe the different subtasks are grouped together. The rules are separated into subtasks both for easier maintenance and to increase efficiency because the interpreter need only consider the rules in a single subtask at any given time.

An example of an XCON rule translated from OPS5 into normal English follows:

R1-Panel-Space

If: The most current active context is panel-space for module-x
And module-x has a line-type and requires cabling
And the cabling that module-x requires is to a panel
And there is space available for a panel in the current cabinet
And there is no panel already assigned to the current cabinet with space for module-x
And there is no partial configuration relating module-x to available panel space

Then: Mark the panel space in the cabinet as used
And assign it the same line type as module-x
And create a partial configuration relating module-x to the panel space.

XCON runs in batch mode, processing an order every minute or two. XCON looks into its database for an order to configure and if it finds an order, XCON configures it, updates the database with its output, and looks for another order.

- *Testing*

At the beginning, Digital used to create a test set of orders, either 25 customer orders or sample orders generated internally. The idea was to test the vast majority of rules on the most difficult orders. Once XCON could run all of these, the developers were confident that the rules functioned well. As new orders came through from customers, the develop-

120 *Getting Started with Expert Systems: A Case Study*

```
COMPONENTS ORDERED
LINE  QTY   NAME           DESCRIPTION                    COMMENT
 1     1    861CB-AJ       8600 QK001-UZ 12MB 240/50HOS
            KA86-AD        PROCESSOR
            12288 KILOBYTES OF MEMORY
 2     1    TU81-AB        1600/6250 BPI 25/75 IPS 240V
 3     2    CI780-AB       780 INTERPROC BUS ADAPTOR 24   1 OF THESE WERE NOT
                                                           CONFIGURED
 4     2    DR780-FB       DMA CHANNEL, VAX-11/780,120V   1 OF THESE WERE NOT
                                                           CONFIGURED
 5     1    DB86-AA        8600 SECOND SBI ADAPTER
 6     1    H9652-FB       8600 UNI EXP CAB 1 BA11A 240
 7     1    RUA60-CD       RA60-CD, UDA50 CTL, W/CAB
 8     3    RA60-CD        RA60-AA, H9642-AR, 50HZ
 9     5    RA60-AA        205 MB DISK, 50/60HZ, NO CAB
10     1    DD11-DK        DD11-D 2-SU FOR BA11-K
11     1    LA100-BA       KSR TERM W/TRACTOR US/120V
12     1    QK001-HM       VAX/VMS UPD 16MT9

COMPONENTS ADDED
LINE  QTY   NAME           DESCRIPTION                    COMMENT
13     1    DW780-MB       8600 SECOND UNIBUS ADAPTER     NEEDED BY THE UNIBUS
                                                           MODULES
14     1    H9652-CB       8600 SBI EXP CAB SWHB 240V3P   NEEDED TO PROVIDE SPACE
                                                           FOR ADAPTORS
15     1    HSC5X-BA       DISK DATA CHANNEL SUP 4 DISK   NEEDED FOR A RA60-AA*

ERROR WARNING

**** THIS NON-STD ORDER REQUIRES MGMT APPROVAL. THERE ARE MISSING MENU ITEM(S)
FROM THE FOLLOWING MENU(S): LOAD-DEVICE


         CABINET LAYOUT


         |------------|
         |CONSOLE     |
         |LA100-BA  0 |
         |            |
         |            |
         |            |
         |            |
         |------------|


|------------|------------|  |------------|------------|
|70-19218-01 |70-19219-01 |  |H9652-CB  1 |H9652-FB  1 |
|FEC CAB # 0 |CPU/KA86    |  |SBI CAB # 1 |UEC CAB # 1 |
||----------||            |  |            |            |
||RL02-FK  0||            |  |            |            |
||          ||            |  |            |            |
||----------||            |  |            |            |
|            |            |  |            |            |
|            |            |  |            |            |
||----------||            |  |            ||----------||
||BA11-AL  1||            |  |            ||BA11-AM  1||
||UBA     0 ||            |  |            ||UBA     1 ||
||----------||            |  |            ||----------||
|            |            |  |            |            |
|            |            |  |            |            |
|------------|------------|  |------------|------------|


|------------| |------------| |------------| |------------| |------------|
|H9642-AR  1 | |H9642-AR  2 | |H9642-AR  3 | |H9642-AR  4 | |TU81-AB   1 |
||----------|| ||----------|| ||----------|| ||----------|| |            |
||RA60-AA  1|| ||RA60-AA  1|| ||RA60-AA  1|| ||RA60-AA  1|| |            |
|'          || ||          || ||          || ||          || |            |
||----------|| ||----------|| ||----------|| ||----------|| |            |
||RA60-AA   || ||RA60-AA   || ||RA60-AA   || ||RA60-AA   || |            |
||UNIT # 4  || ||UNIT # 3  || ||UNIT # 2  || ||UNIT # 1  || |            |
||----------|| ||----------|| ||----------|| ||----------|| |            |
|            | |            | |            | ||RA60-AA   || |            |
|            | |            | |            | ||UNIT # 0  || |            |
|------------| |------------| |------------| ||----------|| |------------|
```

This is output from XCON.

ers would then see where XCON had failed and add those problems to the test cases, constantly making the tests tougher and tougher. These test cases were run against each rule change and/or each formal release. Now, as developers change rules in XCON, they run only those tests they believe are necessary. But before each new release of updated software to the production environment, a complete set of regression tests is run.

- *Integrating the System into the Organization*

The people managing XCON's integration into the organization felt that it was important to minimize the disruption to existing processes. They also understood the importance of facilitating the system's continuing development in order to extend XCON to configure other systems and to be able to add to or modify the system's knowledge.

XCON was first introduced at a final assembly and test plant in Salem, New Hampshire, in January 1980. The technical editors who had been performing XCON's configuration task became XCON's supervisors, examining the output diagrams for correctness.

Today, XCON, running in the Salem plant, services most of Digital's domestic operation. Digital's plants in Ireland and Scotland are also beginning to use XCON.

Initially, a Digital engineer fixed database problems as they became apparent and McDermott fixed rule problems on a monthly basis. Gradually, Digital's internal development team took over the development and maintenance task.

Digital's development team performs the whole ongoing development process, using configuration engineers as consultants. The developers collect and validate new knowledge about the configuration process and about new components. They also encode that knowledge as OPS5 production rules and as input to the component database.

The XCON development team has created a problem reporting system to keep the technicians and the technical editors in constant communication. The system is based on a problem report form for the technicians supervising XCON in the final assembly and test plants to use when they believe XCON has incorrectly configured a system. The problems can now be directed to people with specific expertise.

- *XCON's Performance*

Success for XCON has always been difficult to define. At the beginning, the development team had long and heated debates about the defining criteria. They decided that XCON would have to examine all orders, including the most difficult ones. The degree of XCON's accuracy, as judged by human experts, was initially 75 percent and rose toward a goal of 95 percent over a period of about a year and a half. To increase accuracy was quite a difficult task because the development team was constantly adding new products and finding more and more "hidden" details about how to properly configure a VAX system. Success also became hard to define because the experts often disagreed on what was correct and what was not.

Another measure of success was acceptance by the technical editors and the engineers in the factory. The technical editors and engineers were at first unwilling to accept XCON as a software product. Later, after the development team had run a large number of orders through XCON, the technical editors and engineers accepted the fact that the system worked.

The average runtime required to configure an order is currently 2.3 minutes. Small PDP-11s sometimes take less than 1 minute to configure, while a 200 line item VAX 8600 cluster may take many minutes. Before XCON and before the complications of clusters, technical editors required 25-35 minutes for an average order and got 70 percent of them correct. XCON provides several hundred pieces of information per order, and creates a usable configuration 98 percent of the time. XCON performs about six times as many functions as the technical editors used to perform.

- *XCON's Impact*

XCON has allowed Digital to avoid costs that it would have incurred if Digital had been forced to hire more technical editors as the volume of systems sold increased.

XCON has also made possible another significant cost-avoidance and efficiency measure in the manufacturing process. Before XCON's accurate configuration plans were available, systems were sometimes assembled up to a point at which a problem was discovered, and then the system had to be dismantled and reconfigured. This wasted floor space in the assembly plants as well as time. With XCON, the configurations are dependable, so the manufacturing process can, in turn, be more efficient.

In addition to solving the configuration problem for the manufacturing organization, an expert configurer is important for providing support for the sales force. The configuration exercise defines the actual numbers of each option, the exact power, packaging, and interconnect requirements—information needed in identifying the boxes, cables, terminators, cabinets, and power supplies that must be included in a customer order. Because these elements must be taken into account to provide an accurate quote, the configuration problem must be dealt with at the time a system is ordered. This initially puts the configuration problem on the shoulders of salespeople, who are instead supposed to specialize in dealing with the customers.

Thus, support for the configuration process has meant more efficiency for Digital sales and manufacturing people, and system orders that are complete and functional and therefore cause no delays for customers.

The attitude toward expert systems at Digital has changed from skepticism to confidence, even enthusiasm. Digital is now in the process of developing many more expert systems and AI-based applications. Among them

AI Spear: Computer System Failure Analysis Tool—is an expert rule-based program for analyzing failures in model TU78 tape drives. Spear can detect intermittent failures by analyzing the system error log. Such failures are often impossible to detect with conventional diagnostic programs.

KBTA: (Knowledge Based Test Assistant)—is an expert system designed to aid hardware test engineers in the process of test program development. KBTA implements a complete test programming environment from the acquisition of device information to the synthesis of test programs.

IBUS (Intelligent Business System)—referred to as the IBUS concept, this project is made up of two parts, ILOG (Intelligent LOGistics assistant), on the one hand, and people, on the other. ILOG helps with the execution of orders and order flow, particularly in the handling of problems in manufacturing. The people part of the IBUS concept is concerned with teaching people how to manage change and how to make the focus of the change people, rather than software.

IDT (the Intelligent Diagnostic Tool) — is a program that assists technicians in diagnosing systems that do not pass manufacturing verification tests. It uses search techniques and a representation based on propositional logic to locate failed modules within a system.

INET (Intelligent Network) — is a decision support system that is used for corporate distribution. It is a frame-based simulation system that has knowledge about factories, warehouses, distribution centers, and how to move material from one to another.

ISA (Intelligent Scheduling Assistant) — schedules shop floor activity and manufacturing activity.

NTC (Network Troubleshooting Consultant) — is a rule-based expert system that gives interpretive analysis of and consultative advice on Ethernet/DECnet related problems.

TVX (The VMS Expert) — is an expert system designed to assist users of the VMS operating system. TVX tutors users about the utilities, commands, and functions that the system provides, diagnoses error conditions that the user may encounter, and helps to tailor and structure sets of commands to convert the user's stated goals into VMS primitives.

XSEL (expert SELling tool) — is the adjunct to XCON that helps the sales force interactively configure computer systems to prepare correct quotes right in the customer's office and better match specific products to customer need. XSEL can also submit the orders to XCON in batch mode and generate a complete configuration. XSEL is now running and is beginning to be used by trained sales people in most regions of the United States.

XSITE (expert SITE preparation tool) — is intended to act as an expert site planning assistant, helping in the layout and design of a computer room. XSITE is run after the components are selected, using XSEL.

- *Lessons Learned in Developing XCON*

XCON survived the difficulties of development because of many factors, including the support of management. XCON earned this support by showing enough progress to satisfy management at various time intervals. Another factor in XCON's success was that the people in Digital who were committed to making XCON work did a good job of managing people's expectations. Also contributing to XCON's survival and success was the fact that XCON addressed a problem that desperately needed a solution. And finally, XCON's task also had the right degree of difficulty.

The development of XCON demonstrated the absolute necessity for management support through the trials, tribulations, and pitfalls of getting the program started. AI and expert systems are new and often threatening to the current "installed base" of human experts. Because the technology is new, we do not know exactly what path to follow in all cases and have to learn to "fall forward," so that we can learn from our mistakes and continue to make progress. Management has to be willing to let a development team explore and see what they can and cannot do with the technology.

An expert system needs "mentors" or "gatekeepers" to evaluate its output. These mentors should be experts who can verify the system's output. They need to know what the system is supposed to do and have some understanding of the high-level strategies of the system if they are going to be able to report actual errors and not "bogus problems," that is, problems the system was not designed to solve.

Professor McDermott cites two important lessons that emerged in the development of XCON:

"Recognition knowledge can be used to drive an expert system's behavior, provided that it is possible to determine locally (i.e. at each step) whether taking some particular action is consistent with acceptable performance of the task." and "When an expert system is implemented as a production system, the job of refining and extending the system's knowledge is quite easy."[1]

Some developers on the XCON team feel that, while refining and extending an expert system's knowledge is not exactly "easy," it is amazing that refinement and extension are possible at all. XCON has continued to perform successfully during years of incremental growth and development in the production rule base.

As a result of working on XCON, the developers have recognized two salient points. First, that the typical expert system is never "completed" because the knowledge-intensive problems that it is best at solving constantly change. Second, that the process of ongoing modification of the system's knowledge base is likely to lead to a juncture at which either the whole base or some part of it will need to be redesigned to eliminate redundancies and to make it more efficient and easy to read and maintain.

Getting Started with Expert Systems

This section addresses general issues that arise when you venture into expert systems solutions. The ideas in this section were contributed by, or are supported by, people in Digital who have been involved with expert systems development. The ideas should be construed as conforming to Digital's internal experience, rather than as universally applicable precepts.

When Should an Expert Systems Approach Be Used?

As noted above, expert systems are appropriate for knowledge-intensive problems—problems that we usually turn over to an expert or group of specialists to solve. While expert systems can solve problems that have a single, clear-cut answer, they are also good at solving problems that have more than one correct or acceptable answer. They are skillful at interweaving and balancing many considerations to arrive at some reasonable solution.

Expert systems utilize knowledge and problem-solving skills contributed by experts in a specific problem area. Because such systems depend on human experts for their knowledge, if your experts are currently unable to solve the problem, you won't be able to develop an expert system to solve it, either.

If the difficulty your organization faces is simply that the demand for experts has outstripped their availability, an expert system may be the answer, relieving the bottleneck.

A good problem for an expert system is one in which there is agreement on the facts of the problem domain and in which you can determine some clear definition of the boundaries of your chosen problem area.

Like experts, expert systems don't always arrive at the best possible solution. An expert system will sometimes generate better solutions than an expert, and sometimes solutions that are not as good. In fact, because expert systems can make mistakes, the solutions should be monitored by the user.

- *The Expert System Development Process*

Below is a traditional, generic structure for solving a problem for which an algorithm is known. It is useful for static problems. However, systems designed this way are not usable until they are completed, and if the problem changes, they must be revised or even scrapped. For a typically complex expert system problem, however, the design specifications can't be written without reference to a system that has already been implemented, allowing the implications of various design decisions to become evident. The traditional model must be supplemented for these problems.

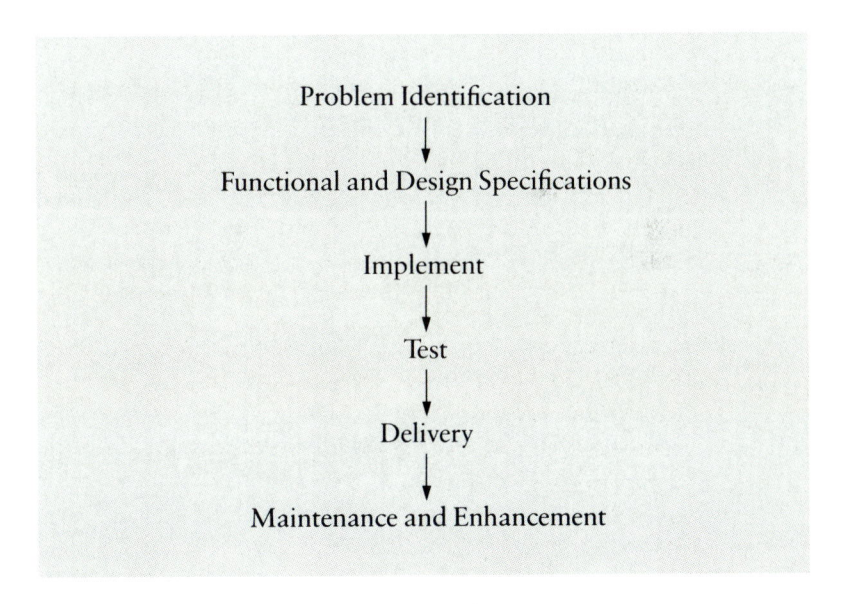

Problem Identification
↓
Functional and Design Specifications
↓
Implement
↓
Test
↓
Delivery
↓
Maintenance and Enhancement

You may be familiar with this sequence in traditional software deelopment.

The following model depicts expert system development. It is useful when no solution algorithm is known. Like the traditional model, the sequence of development phases begins with *problem identification*. In the case of XCON, formal functional specifications were understood rather than written. Because the functional specifications tended to be subject to ongoing revision, it was not practical to retain and reference them.

The next phase of expert system application development—create a *bounded prototype*—supplements the third phase of the traditional cycle, create a *design specification*. One reason for using an AI approach is that, as we noted above, no practical algorithm for solving the problem is known. In that case, the developer usually can't write the design specification demanded by the traditional model. Instead, in the expert system model, when the problem has no known algorithm and you can't write specifications, the developer begins building a bounded prototype in an AI language like LISP. An algorithm may emerge from the bounded prototype, at which point the developer can possibly gain advantages by using the traditional software development model for that piece of the problem. Also in contrast to a traditional system, the bounded prototype, which supplies a solution to part of the problem, can be delivered early in the process and is often useful even at that stage.

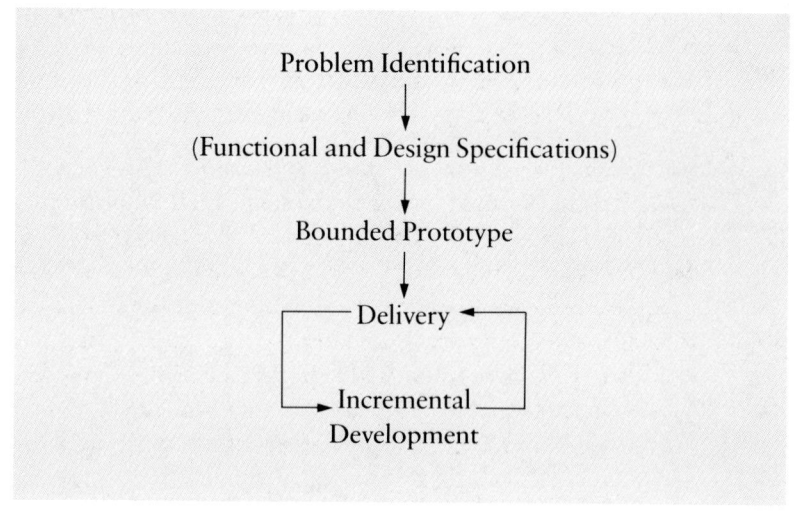

Problem Identification
↓
(Functional and Design Specifications)
↓
Bounded Prototype
↓
Delivery ←
↓
Incremental
Development

A model for expert system development.

Arrows are shown cycling between delivery and incremental development in the expert system development model because this style of system development is suited to dynamic problems. The domains of typical expert systems tend to be subject to rapid change. As we have noted, these problem areas are usually difficult to specify and develop via traditional methods because the problem usually changes significantly prior to implementation. Expert systems allow rapid prototyping and incremental development, which allows the developer to keep up with the evolving problem. Incremental development encompasses the ongoing revision necessitated by the growth of knowledge in a field as well as changes to the task definition. This ongoing revision is similar in function to the entire traditional software development sequence. It is more than simple "maintenance."

- *Organizational Groundwork*

    Generating and Maintaining Support Before embarking on the expert system development process, it is very important to secure the wholehearted support of upper-level managers. The first expert system should be viewed more as an investment in technology and training, early in its development cycle, with the return-on-investment coming later. After a successful introduction of artificial intelligence, subsequent systems can concentrate more on financial return.

    It is very important to manage the expectations people have of the system. A developer should not promise more than can be delivered. It is wise to deliver good-quality, accurate, useful versions of the system early and often, even if the versions deal only with a very small subset of the eventual scope of the project.

    The users should "own" the system. Enid Mumford, Professor of Organizational Behavior, Manchester Business School, discusses this concept of participative design in *Designing Human Systems*. If the users are involved from the very beginning, they can contribute their insights to the system's development, and they are more likely to invest themselves in running the system efficiently. Mumford stresses the importance of allowing people to influence the design of their work situations, both for their own job satisfaction and for efficiency gains.

    Human Resources Required in the Development Effort An expert system development team needs a variety of highly skilled people:

- One or more knowledge engineers. There will probably be a critical shortage of these specialists for the next decade, so it will probably be necessary to train your own knowledge engineers. With training in expert system development (available from universities, systems vendors, and from organizations like Carnegie Group, Inc., Smart Systems Technology, and Teknowledge, Inc.) and the ongoing leadership of an AI consultant, inhouse software talent often works out well. An added benefit is that they know your business and care about profitability. The creative programmers on your MIS team may have the special abilities necessary to be good knowledge engineers. These people should be highly competent in programming, systems analysis, and interpersonal skills and should be willing to learn new problem-solving approaches, languages, and tools.
- Knowledge engineers must be good at communication and general people-skills because they will be working closely with the domain experts for a long time to elicit and articulate the knowledge needed to solve the problem. The knowledge engineer must also understand the needs of the users, because they often are not the same people as the domain experts.
- An AI consultant, if you are not able to directly hire an experienced knowledge engineer.
- Senior technical management, committed managers who can solve problems creatively as they arise and stay with the task for an extended period, if necessary, perhaps a few years.
- Good programmers who can write code to merge the system with interfaces and adjunct systems and who can write traditional computational and database routines that may be used by the AI program.
- Most important, you need experts who are willing and able to articulate knowledge about the domain of the application throughout the life of the system. Experts are necessary during the ongoing revision phase, as well as during original development.

Organizing the Development and Maintenance Team Maintaining an expert system often consists primarily in continuing development and is often performed by the same team. Everyone in the group should have a thorough knowledge of the internal workings of the system. Some of the people continuing development of the system also must be and will, in the course of working with the system, become experts in the domain as well as being creative design level programmers: they need to possess the skills of the knowledge engineer. This allows them to work directly with the domain experts.

Funding for the project should be independent of short term considerations, goals, and budgets because expert system development may require a longer term.

- *Problem Identification*

There are said to be fewer than 1,000 AI researchers in the world today who can recognize where an expert system could be useful and then implement it. If you can't hire one of these scarce experts, an alternative is to send one of your organization's engineers off to acquire AI knowledge. The engineer would also spend some time consulting with a knowledge engineer. The process could take years of reading and investigating. However, if the engineer spends some of that time in training with one of the firms that specialize in teaching AI techniques, the time could be shortened. Currently, the demand for the services of these firms is about equal to the supply.

Once the problem has been identified, it must be bounded. In order to do this, a knowledge engineer examines the subject matter of the problem domain, identifying the domain's components, the relevant data, and its necessary functions and interactions.

At the outset, don't specify the system too rigidly. Leave room to make ad hoc decisions. The functionality of the final system may be different from, but equally valuable as, the originally projected system. The task should be able to bear redefinition as system performance changes. The approach and structure of the system will also change. Nevertheless, each time you articulate or revise the goals of the system, plan tests that will measure success so that the goals, even if they change, are clear to the developers.

- *Encoding the Knowledge*

Before beginning work on a particular expert system, the knowledge engineer does background reading on the problem domain. This helps the knowledge engineer understand the expert's vocabulary and provides him or her with a conceptual map of the domain.

Then the knowledge engineer talks with the expert about what knowledge and skills are used to solve the chosen problem. It is also beneficial for the knowledge engineer to watch the expert performing the task. It can be beneficial to have more than one expert in order to highlight and then resolve inconsistencies. The knowledge engineer should ensure that the experts do not trim their knowledge to fit the knowledge structure more conveniently; it is important to retain the nuances of the knowledge. In addition, the experts should be continuously, not intermittently, available both to ensure that the knowledge engineer does not have to take on the expert's role and to provide good test examples throughout the development process. While the interaction between the consultant and the team is not constant, access to the consultant must be immediate.

Working and consulting with the experts, the knowledge engineer structures the problem and chooses a means of knowledge representation and the appropriate inference mechanisms, encodes the knowledge, verifies the knowledge (by reviewing it with the experts), and then modifies the knowledge as required.

The initial knowledge base is likely to require revision, the need for which becomes apparent when the system is tested. At this point, special cases and exceptions that were not covered originally may become apparent, giving the expert an opportunity to articulate the specialized knowledge.

During this knowledge acquisition process, a new way of seeing the problem may emerge and suggest a revision of the initial choices of techniques, the knowledge representation, and/or the inference mechanism. The scope of the underlying problem often changes. For example, major technology and architectural revisions were required in order to add to XCON the ability to configure clusters and networks. A revision may also become necessary if the knowledge base grows too large to work efficiently in the original framework.

- *Prototype Design and Development*
  A good prototype has been described as "small enough to implement, large enough to impress, cheap enough to get funded, and urgent enough to get commitment."

A good task for the prototype is one that can be performed using a subset of the full knowledge base, so you don't have to engineer the full domain before seeing how feasible the approach is.

The prototype should define a representative subset of the problem, preferably a subset that can make a worthwhile contribution in order to confirm the system's utility. If the prototype focuses on a subset of the complete problem that benefits the domain expert, he or she is likely to stay positively involved in the project. The prototype may be useful to the expert early on, even before all of the goals of the final system have been met. It can even enhance the expert's functioning by performing some of the more mundane tasks.

After the boundaries of the prototype have been determined, the developers set preliminary functional specifications in accordance with the users' needs. The prototype should incorporate the intended interface, so users can see how comfortable and efficient it is.

The knowledge engineer then selects an appropriate form of representation, elicits the knowledge from the expert, and fits the knowledge of the problem domain into the knowledge structure. The knowledge engineer also creates an initial design that will incorporate the knowledge base and the chosen AI methods and inferencing mechanisms.

The prototype system must be delivered to the users as early as possible. Rapid prototyping helps to show the feasibility and the appropriateness of the chosen form of implementation early on. Demonstration of the prototype also allows the users to indicate any modifications needed in the final design. From this early delivery on, new functionality is added incrementally until the problem is fully solved.

- *Testing*

Tests should relate to and verify the performance goals and acceptance criteria defined earlier in the process. The best tests consist of what the experts think are the most revealing, tough, and sensitive problems. It is good to get feedback on a very large volume of real, as well as test, runs of the system by using it in parallel with manual techniques.

Cooperative domain experts are important because they are needed to provide feedback on test runs of the system. Cooperative users are important, as well, because they are needed to provide continuous feedback on real runs of the system. This feedback is needed early in the development process to help the system grow and later to maintain a high degree of integrity.

Not much is known yet about testing expert systems. Large expert systems may, for all practical purposes, have an infinite number of solution paths, and it is impossible to test all of them. For XCON, hundreds of regression tests are run before each release. As many runs as possible should be made and evaluated before a release. An attempt should be made to exercise the major paths in the system, as well as running a sampling of typical tests.

It is helpful to have experts, especially external consultants, assess prototype performance. For systems that include an explanation facility, the expert should evaluate the system's explication of the solution path in addition to the solution itself to make sure it reflects reality and is both succinct and sufficient.

- *Delivery*

Early delivery of the prototype makes it possible to refine the system through trial use. Several attempts may be required to transform the prototype into a fully functioning system.

It takes good planning to introduce the system to users, and attitudes and change must be managed with sensitivity. As mentioned earlier, a participative design approach, in which it is understood that the *users* own the system, contributes to the efficiency of, and receptivity to, the system as well as to the job satisfaction of the users. If possible, the system should be understood as an aid or tool and not as a replacement for people performing the task. Focusing on the capabilities of the system, rather than on the "new technology," makes the system less intimidating. Make it clear that current jobs are to be changed, not replaced, whenever possible. Great sensitivity is needed to manage the issue of alterations in jobs.

Like experts, expert systems don't necessarily come to the best possible solution. Their purpose is, like the expert's purpose, to come to a good enough solution relative to the time and resources expended. Moreover, an expert system should be expected, like a human expert, to make mistakes. Knowledge is a moving target, never exhaustively comprehended, especially for the volatile problem domains expert systems can help people manage. People who use an expert system must be prepared for a system that has imperfections. It should also be remembered that an expert system is still a software system, vulnerable to the usual problems.

- *Incremental Development to Deal with Growth and Change*
An expert system is never "done." Development is likely to continue as the application is integrated into your operations because the problem definition will change, as will the data the system works on and knowledge about the problem domain. Incremental development requires a continuing relationship with the domain expert. Testing must also accompany changes in the knowledge base.

Continuing development can be the most expensive aspect of an expert system, but it is also what keeps the system from becoming obsolete. If the system is large and complex, and if there is a need for other AI projects, it may be worthwhile to establish or purchase a formal training program. Developing a training program can be an especially powerful way to utilize the knowledge of AI consultants.

Resources to Aid You in Getting Started

You may be able to use an off-the-shelf knowledge engineering tool to structure and input knowledge for a system. A good one

- Supports a convenient language in which to express the knowledge.
- Supplies a compiler for this language.
- Provides the software necessary for the smooth creation and running of the expert system.
- Provides aids for testing, verifying, and tuning the system.
- Provides facilities to make maintenance and modification convenient.

Many knowledge engineering tools have been derived from successful expert systems. The knowledge bases have been extracted, leaving behind the structural and inferential aspects of the system, sometimes referred to as "shells." Shells can contain generic knowledge of a task. Another similar tool, called a programming environment, provides a set of facilities, such as an editor, interactive debugging facilities, and input/output routines to aid in the various phases of program development. Some AI languages, like LISP, have enough facilities to be considered environments on their own.

Different knowledge engineering paradigms, or strategies, as well as inferencing methods and other features are reflected in these tools. The following list describes some of the better-known, commercially available languages and tools. The list will give you an idea of the scope and size of the field. However, products are changing fast. By the time you read this book, the list will probably be out of date.

- VAX LISP, an implementation of COMMON LISP, running on VAX superminicomputers under the VMS operating system. Features include availability to the user of both interpreter and compiler modes, dynamic linking of compiled and interpreted code, lexically scoped variables, a user-extensible editor and user-controllable utility for enhancing printer readability, and the facility for calling routines written in other languages in the VMS environment.
- GCLISP (GOLDEN COMMON LISP) developed by Gold Hill Computers, Inc., Cambridge, Massachusetts, and available from Digital Equipment Corporation, runs on Digital's Rainbow 100 and Rainbow 100+ computers under the MS†/DOS operating system. GCLISP is ideal for entry-level AI operation. The program package was designed to provide training for programmers as well as for program development. As a subset of COMMON LISP, it is compatible with Digital's VAX LISP.
- INTERLISP for VAX was developed by the Information Sciences Institute (ISI) at the University of Southern California and is available from Digital Equipment Corporation. INTERLISP is a fully integrated set of programming tools coupled with a complete implementation of the LISP language. INTERLISP's integrated program support facilities include syntax extensions, uniform error handling, automatic error correction, an integrated structure-based editor, a sophisticated debugger, a compiler, and a filing system. INTERLISP runs under VMS and UNIX BSD 4.2 operating systems. Both implementations are from ISI.
- VAX OPS5 is available from Digital Equipment Corporation. VAX OPS5 facilitates the expression of knowledge in "If...then...." production rule form. The language operates on a recognize-act cycle, and it has a built-in conflict resolution strategy that chooses one from among the rules that match the data in working memory. VAX OPS5 is a highly efficient general purpose expert systems language and has an easy-to-program interface to all other languages available on the VMS operating system.

- Quintus Prolog, developed by Quintus Computer Systems, Inc., Berkeley, California, runs on VAX computers under VMS and UNIX operating systems, as well as other machines.
- Prolog II, developed by Prologia, Marseilles, and available in Europe from Digital Equipment Corporation, runs on the VAX superminicomputer and on MicroVAX workstations under the VMS operating system and under the MS/DOS operating system on Digital's Rainbow 100 personal computer.
- Knowledge Craft,† from Carnegie Group, Inc., Pittsburgh, is an integrated knowledge engineering and problem solving environment for building large knowledge-based systems. Knowledge Craft was implemented using SRL (Schema Representation Language) developed by Carnegie-Mellon University. Important features of Knowledge Craft include OPS5 and Prolog programming "workbenches," knowledge-based editors, powerful graphics facilities, and a database management system.
- Language Craft,† also from Carnegie Group, Inc., is an integrated natural language environment that provides an expandable software system for developing natural language interfaces. You can use it to construct general and application-specific language interfaces to databases, operating systems, expert systems, or commercial software applications.
- ART† (Automated Reasoning Tool), from Inference† Corporation, Los Angeles, runs on Digital Equipment Corporation VAX computers and on Symbolics and LISP Machine Inc. LISP machines. ART incorporates a flexible, generic paradigm including frames, rules and processes, many internally available AI algorithms and data structures, and all of the major AI problem-solving techniques. The resulting code is in LISP.
- KEE† (The Knowledge Engineering Environment for Industry) is a knowledge-engineering environment developed by IntelliCorp, Menlo Park, California. KEE incorporates frame-based and rule-based reasoning paradigms, knowledge manipulation, reasoning, and object-oriented programming. Interactive graphics model the system's knowledge structure, so the user can visualize it.
- LOOPS is an extension of the INTERLISP-D programming environment for Xerox† 1100 series scientific information processors. LOOPS combines several programming paradigms: procedure-oriented, object-oriented, data-oriented, and rule-oriented paradigms. LOOPS has

access-oriented programming, meaning it uses on-screen gauges to monitor effects of changes of data.
- DUCK,† from Smart Systems Technology, McLean, Virginia, produces LISP code, incorporates several programming paradigms—logic programming, rule-based systems, nonmonotonic reasoning, and deductive forward and backward search. DUCK is useful for building intelligent databases and knowledge-based expert systems.
- EXPERT-EASE,† from Jeffrey Perrone & Associates, Inc., San Francisco, runs on the IBM Personal Computer or PC XT and some compatibles. EXPERT-EASE allows the user to create models of expertise that can then be rapidly converted into inquiry systems. EXPERT-EASE is suited to building expert systems for small-to moderate-sized tasks.
- S.1† from Teknowledge, Inc., Palo Alto, California, is a complete package of software, training, and support for experienced knowledge engineers to use in developing large-scale practical applications. S.1's interactive graphics facility dynamically displays an event tree that illustrates the flow of a consultation.
- TIMM† (The Intelligent Machine Model), from General Research Corporation, McLean, Virginia, uses a frame paradigm and a partial match analogical inferencing procedure so that the system can make use of incomplete and approximate knowledge. TIMM builds expert systems from representative examples of the type of decision being modeled, eliminating the need for the explicit stating of rules.

The following tools were developed in the research community. The availability for commericial use of some of them may be limited:

- AGE, developed at Stanford University, is based on LISP and can build various problem-solving architectures.
- EMYCIN, a production-rule based structure developed at Stanford, is appropriate for developing a system that produces analyses or interpretations. The user supplies data in response to requests from the system. This skeleton expert system was derived from MYCIN, a system for diagnosing meningitis and blood infections. EMYCIN uses a backward-chaining control strategy. Facilities include an explanation program, and tracing and debugging facilities.
- HEARSAY III comes from the speech understanding systems HEARSAY I and HEARSAY II, developed at Carnegie-Mellon University (CMU). HEARSAY III can bring diverse knowledge domains to bear on a partic-

ular application domain. HEARSAY III uses a blackboard, a database that allows processes and cooperating expert subsystems to communicate with each other.
- KAS (Knowledge Acquisition System) is a knowledge engineering tool for the PROSPECTOR expert system, developed at SRI International. This system can accommodate both forward and backward chaining, and it has a convenient knowledge-based editor. KAS is appropriate for developing diagnosis and explanation systems.
- OPS4 and OPS5, developed at CMU, are useful for building production systems. OPS5 has a built-in inference engine and an automatic pattern matcher. The commercial implementation VAX OPS5 is notable for its speed.
- ROSIE,† developed by Rand Corporation, is a general purpose, rule-based, procedure-oriented system.
- SRL (Schema Representation Language) was developed at CMU by Dr. Mark S. Fox and J.M. Wright in the late 1970s. SRL is a frame-based language, written in LISP, running in a LISP environment. SRL gives the user a set of predefined relations, such as IS-A and INSTANCE. The definitions of IS-A and INSTANCE implement inheritance using their slot values as the way to get values of particular slots from other frames. CMU's implementation of SRL comes with several editors for schemata, a command interpreter that provides help, a graphics package, and a window package. The graphics package is capable of translating to graphic images from schemata.

- *Commercially Available Training and Assistance*

Commercial firms have begun to offer training and consulting in artificial intelligence techniques.

Carnegie Group, Inc., Pittsburgh, was founded by members of the Carnegie-Mellon University Computer Science Department. Carnegie Group offers the 1-day overview called "Briefings on Artificial Intelligence," the 5-day program called "The Foundations of Artificial Intelligence" and more advanced hands-on technical courses. Carnegie Group is also developing and customizing AI software tools and systems for its customers.

Digital Equipment Corporation's Educational Services offers training in AI and related languages and Software Services consulting services.

LISP Machine Inc. has a center in Cambridge, Massachusetts, where it offers training in LISP and in the use and maintenance of their LISP machines.

Symbolics, Inc., has training centers in Cambridge, Massachusetts, and San Francisco, California. Symbolics offers a complete set of courses at introductory, intermediate, and advanced levels in Symbolics LISP,† an extended superset of COMMON LISP. Symbolics also offers courses on its hardware and software systems. LISP machines are available for the use of all students in all courses.

Smart Systems Technology, McLean, Virginia, offers an 8-week modular training program in developing expert systems and the use of expert system development software tools, including LISP and OPS5. Smart Systems also offers consulting as a follow-on to the training, as well as AI systems implementation.

Teknowledge, Inc., Palo Alto, California, was founded by members of Stanford University's and MIT's AI departments and people from the Rand Corporation. Teknowledge develops expert systems for customers. Teknowledge also provides courses on the use of their knowledge engineering software products and training ranging from 1-day seminars to 6-month courses in techniques of knowledge engineering.

In addition, training companies and research firms have begun to run seminars and forums on various aspects of AI in general and expert systems in particular.

- *Resources from Digital Equipment Corporation to Help You Get Started*

Digital Equipment Corporation's Software Services Software Services is actively developing expertise in artificial intelligence methods and technologies to meet the emerging demand for consulting support for pioneering customers, as well as for artificial intelligence software products and support services.

The full range of software product services on artificial intelligence products such as LISP and, in the near future, OPS5, and ART are offered. All Digital-supported artificial intelligence software products are backed up with trained specialists who can help by phone or by onsite visits.

The following consulting services are offered:

- Artificial Intelligence Expert System Application Survey. Software Services offers a service that identifies potential artificial intelligence applications in an organization. A Digital consultant interviews managers and knowledgeable users about the organization's activities and processes to identify candidate artificial intelligence applications and evaluate these candidates against technical and business criteria, including estimating payback potential. From this evaluation, the consultant develops a recommendation and produces a written report of the findings. This survey normally takes 2 to 4 weeks.
- Development of an artificial intelligence Expert System. Software Services will accept these development projects on a limited basis. A growing number of Software Services specialists are being trained in this emerging technology. They are supported by senior knowledge engineers in our corporate Artificial Intelligence Applications Group. Software Services recognizes that the building of an expert system can be an important element in transferring this technology into the customer's organization. We typically plan the project to include frequent reviews and orientation sessions for the customer's technical staff.
- Orientation Sessions. For subjects on artificial intelligence currently unavailable through Digital Educational Services, Software Services can tailor lectures and seminars to meet customers' specific needs. These sessions could include the use of artificial intelligence tools (such as OPS5), an overview of artificial intelligence, and expert system development methodology.

Software Services expects to continue to exand in staff and services in this emerging major technology.

Training Available from Digital's Educational Services Digital Equipment Corporation's Educational Services offers courses and seminars in artificial intelligence, focusing primarily on languages and tools for experienced programmers.

The LISP language course, taught in Digital's Bedford, Massachusetts, Training Center, specifically addresses COMMON LISP implemented on VAX computers. The course, which lasts two weeks, is intended for experienced programmers and analysts who intend to program applications in LISP in commercial and research environments.

Three seminars are also offered—*Introduction to Artificial Intelligence, VAX LISP for Experienced Programmers,* and *An Overview of OPS5. Introduction to Artificial Intelligence* is intended primarily for managers and technicians who want to explore the introduction of artificial intelligence techniques into their current software applications. The seminar focuses on major companies ranging from computer manufacturers to companies in the oil industry that are devoting substantial effort to developing their AI expertise. This 3-day seminar explains the basic concepts and techniques that underlie AI. Beginning with a tutorial on LISP and Prolog, the course will also cover major AI concepts including knowledge representation, problem solving, formal and informal methods of reasoning, and search strategies. In addition, the course takes an in depth look at two areas where AI techniques have been applied successfully in industry—expert systems and natural language systems. Several commercial systems including Digital's XCON will be discussed as well as current trends in the Fifth-Generation Computer Systems Project.

VAX LISP for Experienced Programmers focuses on the LISP language as the single most important tool in artificial intelligence development and research over the past 25 years. The seminar addresses LISP's suitability for symbolic computation and for implementing embedded languages as well as discussing the large number of dialects associated with the language. The seminar emphasizes COMMON LISP as the de facto standard and VAX LISP as the most widely used implementation of COMMON LISP. The approach is at first interactive with many exercises in gradually writing more sophisticated LISP programs. The student learns the fundamental techniques of recursive programming and also gains an appreciation for the applications for which the language is well suited. The seminar is accompanied by an extensive set of notes and the classic text *COMMON LISP: The Language* by Guy Steele.

An *Overview of OPS5* provides a working knowledge of the OPS5 production language optimized for expert systems. It provides an overview of the basic architecture of rule-based systems. Through a series of in-class exercises and demonstrations, the attendees learn to evaluate the appropriateness of OPS5 as a language for their specific application.

Educational Services Seminar Programs also offers two seminars only at customers' sites: *AI for managers,* a 1-day nontechnical overview of the field of artificial intelligence, and *Logic Programming Languages,* an in-depth comparative analysis.

Besides the courses and seminars offered on artificial intelligence, Educational Services offers a large range of supporting courses and seminars on VAX superminicomputers, the VMS operating system, layered products, and traditional languages.

Educational Services also schedules, on a periodic basis, special courses specifically focused on the development of the skills of AI practitioners, such as knowledge engineering. These special courses use workshop techniques to ensure hands-on student interaction with AI development tools. They cover such subjects as how to manage artificial intelligence applications, successful interviewing of an expert in the development of an expert system, and understanding the difference between rule-based languages and EDP languages and tools.

Books Available from Digital Press Related to Artificial Intelligence

COMMON LISP: The Language, by Guy L. Steele, Jr., 1984, paperbound, 480 pages, order number EY-00031-DP, ISBN 0-932376-41-X.

COMMON LISP: The Language is the standard reference for COMMON LISP, the LISP dialect that is rapidly becoming the de facto standard of the LISP programming community. COMMON LISP is the result of three years of collaboration by artificial intelligence researchers and LISP language experts in the academic community, government, and industry. This language is powerful, portable, and stable. Compatible with many older implementations of LISP, COMMON LISP incorporates more features, including a rich and more complex set of data types and control structures.

This book contains many examples of the use of COMMON LISP functions. Implementation notes suggest techniques for handling tricky cases. Compatibility notes compare or contrast COMMON LISP features with those of other widely used LISP dialects.

Engineering Intelligent Systems: Concepts, Theory, and Applications, by Robert M. Glorioso and Fernando C. Colon Osorio, 1980, hardbound, 472 pages, order number EY-AX011-DP, ISBN 0-932376-06-1.

Engineering Intelligent Systems presents the design and application fundamentals of artificial intelligence techniques. This is the first book to tie together current intelligent systems concepts and techniques for such diverse fields as electrical engineering, mathematics, and neurophysiology.

❊ Annotated Bibliography

Bachant, Judith, and John McDermott, "R1 Revisited: Four Years in the Trenches," *AI Magazine*, fall 1984, pp. 21-32. This article describes the process of extending the expert system R1's (XCON's) knowledge base and evaluates the system's performance over a four-year period. The authors share the conclusions from their experiences working on the largest commerical expert system in daily use.

Brownston, Lee, et al., *Programming in OPS5: An Introduction to Rule-Based Programming*, Reading, Addison-Wesley Publishing Company, 1985. This book is for experienced programmers who wish to develop techniques for rule-based programming in OPS5. The first part is a tutorial on OPS5 and related programming techniques. The second part compares OPS5 with other programming tools.

d'Agapeyeff, Alex, "Report to the Alvey Directorate on a Short Survey of Expert Systems in UK Business," *Alvey News*, Supplement to Issue No. 4, April 1984, 11 pp. Principal findings of this report include the ideas that expert systems in business can be much simpler than had been thought necessary; they can produce modest usage gains while still incomplete; they can with care be built by self-taught teams; and expert systems are not inherently complex, risky, and demanding.

Dickson, Edward M., "Comparing Artificial Intelligence and Genetic Engineering: Commercialization Lessons," *AI Magazine*, winter 1985, pp. 44-47. This article compares the emergence of commercial genetic engineering and commerical AI and applies lessons learned in genetic engineering to the commercialization of AI.

"*Expert Systems* Interview: Alex d'Agapeyeff," *Expert Systems*, Vol. 1, No. 2, 1984, pp. 129-135. This article expands on points brought up in d'Agapeyeff's report to the Alvey Directorate, cited in this bibliography.

Feigenbaum, Edward A., "Themes and Case Studies of Knowledge Engineering." In Donald Michie (ed.), *Expert Systems in the Micro-Electronic Age*, 1979, Edinburgh University Press. This book discusses artificial intelligence and knowledge engineering in the context of case studies.

Hayes-Roth, Frederick, "Knowledge-Based Expert Systems," *IEEE Computer*, October 1984, pp. 263-273. This article discusses the current state of knowledge systems technology and its commercialization.

Hayes-Roth, Frederick, "The Knowledge-Based Expert System: A Tutorial," *IEEE Computer*, September 1984, pp. 11-28. This article discusses the uses for expert systems, the state of the art, major developments in the field, scientific and engineering issues, the process of building expert sytems, the role of tools, and interfaces. Hayes-Roth presents an overview of the field of knowledge engineering.

Hayes-Roth, Frederick, Donald A. Waterman, and Douglas B. Lenat, *Building Expert Systems,* Reading, Massachusetts, Addison-Wesley Publishing Company, Inc., 1983. This book is the product of a workshop intended to synthesize knowledge in the expert systems field. It identifies the things an engineer or technical manager needs to know in order to undertake the implementation of an expert system. The book includes comparisons of various knowledge engineering tools and provides in-depth information on the different stages in implementation. Also noted are potential pitfalls and ways to avoid them.

McDermott, John, "R1: A Rule-Based Configurer of Computer Systems," Carnegie-Mellon University Report CS-80-119, April, 1980. This report describes the development of the computer configuration system R1, now known as XCON.

McDermott, John, "R1's Formative Years," *AI Magazine,* Vol. 2, No. 2, summer, 1981, pp. 21-29. McDermott presents a detailed discussion of R1's design and implementation history, highlighting issues of attitudes, of the feasibility of AI approaches, and of expectations for the system.

Mumford, Enid, *Designing Human Systems,* Manchester, England, Manchester Business School, 1983. This book discusses the ETHICS (Effective Technical and Human Implementation of Computer-based Systems) Method for managing technical change.

Polit, Stephen, "R1 and Beyond: Technology Transfer at DEC," *AI Magazine,* winter 1985, pp. 76-78. The author discusses the transfer of AI technology from academia to industry, illustrated by his experiences with Digital Equipment Corporation's R1 (XCON) expert system.

Sagalowicz, Daniel, "Development of an Expert System," *Expert Systems,* Vol. 1, No. 2, 1984, pp. 137-141. This paper presents methods for identifying problems that are appropriate for knowledge systems solutions, ways to assess risks and potentials, and a knowledge-engineering methodology.

Smith, Reid G., "On the Development of Commercial Expert Systems," *AI Magazine,* fall 1984, pp. 61-73. This article describes Schlumberger's experience developing the Dipmeter Advisor expert system for oil-well log interpretation.

Wright, J.M., and Mark S. Fox, *SRL/1.5 User Manual,* Pittsburgh, Carnegie-Mellon University, Robotics Institute, 1983. This is a users' guide to SRL. It includes background on reasons for development of SRL, an overview of features of the language, and instruction in coding SRL.

Note

1. McDermott, John, "R1: A Rule-Based Configurer of Computer Systems," Carnegie-Mellon Report CMU-CS-80-119, April 1980, p. 1.

Chapter 7

National Programs and Other Major Cooperative Efforts in AI Research

It has been suggested that in the coming decades, the greatest economic leverage and world power will be held by those companies and nations that can manage information most effectively. Advanced computing systems and the understanding of how to use them effectively may be the critical factors upon which power rests. In addition to being able to manage information with the greatest efficiency, companies and countries with the best advanced computing technology should be able to dominate the world computer market. The winner of the race to develop advanced computing systems will be able, at least initially, to set standards for the next generation of computers, and other manufacturers will have to alter their designs to be compatible. In national defense, whoever has the more powerful computers and more intelligent systems may be able to control significantly more powerful and precise weapons systems. And greater precision may make it possible to reduce the size of arsenals, cutting their costs.

Advanced computer systems are also expected to directly raise the standard of living in societies with access to them. For example, advances in robotics may make it possible to produce mechanized assistants and improved mobility aids for people with disabilities. Good natural language interfaces may make it possible for more people to have access to sources of information that only the technically educated or wealthy now have. Better management of information may make it possible to refine the marketing process so that production is more perfectly aligned with demand and there is less waste and greater consumer satisfaction. Cars may be equipped with expert diagnostic systems. Individuals may even be able to use inexpensive versions of expert systems that run on personal computers to manage their household energy consumption.

With potentials for advances in the standard of living and with economic and military preeminence at stake, research in advanced computing hardware, software, and techniques is receiving heavy stress around the world. Research programs are under way not only in universities and in individual corporations but also in programs administered by individual national governments, allied governments, and consortia of corporations, some of whom normally compete with each other. These groups are all seeking to be the first to achieve advances leading to the *fifth-generation computer*. The term *fifth-generation* is used in Japan to refer to a non-von Neumann computer system capable of processing inferences. In the United States, the definition of the fifth-generation computer varies, but it usually includes knowledge processing capabilities and a new, parallel architecture that will allow for increases in speed of 100 to 1,000 times over current computers.

Computers currently in use are based on what is known as a von Neumann architecture. Von Neumann computers carry out processes sequentially and are composed of a central processing unit, memory, and input and output devices. The first four generations of computers (chronologically, computers based on electronic vacuum tubes, on transistors, on integrated circuits, and on very large-scale integrated (VLSI) circuits) have been based on the von Neumann architecture. It is believed that sequential computers are approaching the limits of their potential speed. The fifth-generation computer, however, will probably be based on an architecture that can carry out multiple processes at the same time "in parallel" carrying out tasks at greater speed than a sequential machine. It must be noted, however, that some tasks are inherently sequential, depending on the results of previous processes, so a parallel machine will not be the answer for all computing problems.

In addition to developing a fifth-generation computer suitable for artificial intelligence work, research around the world is aimed at developing more powerful supercomputers that would operate at tremendously faster rates than today's most powerful supercomputers. While it is the fifth-generation computers that directly address the needs of artificial intelligence computing, the supercomputers will also be useful for AI in that they will be able to run today's programs much faster.

Japan was the first country to declare its intention to make the fifth-generation breakthrough. This announcement prompted the United

States and several European countries to parallel the Japanese effort with programs of their own.

Japan and the Fifth-Generation Computer Systems Project

Japan's Ministry of International Trade and Industry (MITI) is directing Japan's effort to develop advanced computer technologies. MITI is funding the National Superspeed Computer Project as well as the Fifth-Generation Computer Systems Project.

The Institute for New Generation Computing (ICOT) was formed to carry out the Fifth-Generation Computer Systems Project. ICOT began operations in the summer of 1982. For the Japanese, the fifth-generation computer means a fast, logic-based, knowledge-processing computer with an interface that can understand and respond in natural language. They have decided to base the machine on a parallel architecture and on logic programming utilizing Concurrent Prolog (also called Parallel Prolog), developed by Dr. Ehud Shapiro currently at Weizmann Institute of Science in Israel.

Because Japan is a country of limited natural resources, its government sees winning the race for a fifth-generation computer as the primary hope for the country's economic future. If Japan is able to set the standards for advanced computer systems, its economy will be able to make the most of Japan's most valuable resource—its people's intelligence. Another factor that makes advances in computing critical to Japan is the need for a means to allow Japanese programmers to interact with computers using their everyday language. Currently, the written form of the Japanese language is not amenable to programming because it has so many characters. Therefore, Japanese programmers have to work in foreign languages. The programmers would be able to work much more efficiently if they could work in their own language. The Japanese feel that their fifth-generation effort is a necessity for economic survival.

The Fifth-Generation Computer Systems Project was organized for the government by Tohru Moto-oka, a professor of electrical engineering at Tokyo University, and Dr. Kazuhiro Fuchi, former head of the Information Sciences Division of MITI's Electrotechnical Lab. ICOT intends to develop hardware and software that will facilitate the production of expert systems, natural language systems, and robotics applications. Dr. Fuchi heads the Fifth-Generation Computer Systems Project.

ICOT is a collaboration of eight member companies—Fujitsu, Ltd., Hitachi, Ltd., Matsushita Electrical Industrial Co., Mitsubishi Electric Co., Nippon Electric Corp., Oki Electric Industry Co., Sharp Co., and Toshiba Corp. Nippon Telephone and Telegraph Public Corp. and the Electrotechnical Lab of MITI are partners.

The participating companies have provided ICOT with researchers who work for ICOT for a period of 3 to 4 years, during which time they occasionally report to their companies on the research. At the end of this time, they return to their companies, bringing new expertise with them.

MITI is budgeting $450 million over 10 years for the Fifth-Generation Computer Systems Project, with Japanese industry at least matching that amount. MITI funds are allocated by ICOT, per contract, to the participating companies for work performed.

One project within the Fifth-Generation Computer Systems Project is to develop the Personal Sequential Inference machine (PSI), a Prolog version of a LISP machine. A knowledge programming system called Mandala will be the counterpart to the LOOPS or SMALLTALK environment of a LISP machine. Kernel languages are also being developed that will define interfaces between the hardware and software.

The Superspeed Computer Project aims to produce parallel processing computers that are 100 to 1,000 times as fast as present-day supercomputers. MITI has budgeted about $100 million for this effort over an 8-year period.

Major Programs in the United States

Microelectronics and Computer Technology Corporation
The Microelectronics and Computer Technology Corporation (MCC) was formed in 1982 in order to respond to challenges to U.S. preeminence in technology. Other countries, such as Japan and the United Kingdom, have brought together the efforts and financial support of industry and government in a way that companies in the United States were unlikely to be able to challenge acting individually. However,

instead of an association between industry and government, the organizers of MCC felt that the most effective vehicle for achieving their aims was a consortium of private companies. In MCC these companies pool their resources to create a research and development environment designed to keep the United States foremost in computer technology. MCC's charter is to perform research that will assist the member companies to produce products that are technology leaders.

MCC is an independent for-profit corporation in which the shareholders each hold equal shares. In addition to buying a share in MCC, all member companies must join one or more programs and contribute outstanding scientists and engineers to those programs. The current price for one share of stock is $1 million. Currently, there are four project areas. While all the patents arising from the research belong to MCC, the technologies are licensed for the first 3 years to the companies that produced the technology. After the initial 3 years, the technology will be available to other MCC shareholders and to nonshareholders at the discretion of the MCC board of directors. The annual budget of MCC will be about $65 million in 1986.

As of March 1985, there were 21 member companies. The current member companies are Advanced Micro Devices, Inc.; Allied Corp.; Bell Communications Research; BMC Industries, Inc.; Boeing Corporation; Control Data Corp.; Digital Equipment Corp.; Eastman Kodak; Gould Inc.; Harris Corp.; Honeywell, Inc.; Lockheed, Inc.; 3M Corp.; Martin-Marietta Corp.; Mostek Corp.; Motorola, Inc.; NCR; National Semiconductor; RCA Corp.; Rockwell International; and Sperry.

MCC got under way in its Austin, Texas location in 1983 under the leadership of its president, Admiral Bobby Ray Inman (retired), former director of the National Security Agency and former deputy director of the Central Intelligence Agency.

The four technical project areas are

1. Advanced Computer Architecture, aimed at developing advanced technology in four areas:

- Human Interface—to explore all aspects of interaction between people and computers with emphasis on ease of learning and increased functionality through new technology.

- Artificial Intelligence—to expand knowledge of expert system design and implementation to support construction of complex production-level expert systems in a timely and efficient manner.
- Parallel Processing—to achieve much higher performance at cost-performance levels comparable to microcomputers. Improved reliability, modularity, and programmability are related goals.
- Database—to build hardware and software systems that yield high performance in processing complex queries across very large databases, and that augment the programming of data access.

2. Microelectronic Packaging and Interconnect

3. Advanced Software Technology

4. CAD for VLSI

These programs will develop proprietary designs that can be adapted for marketable products by the companies sponsoring the research.

Technology developed at MCC is transferred back to participating companies by formal documentation, technical presentations, by the return of employees who have carried out their assigned tenures at MCC, as well as by communication from liaison representatives from each company, who reside at MCC and report results of the research to their companies on a regular basis.

- *Programs of the Defense Advanced Research Projects Agency*

The United States government's Defense Advanced Research Projects Agency (DARPA) has heavily supported artificial intelligence research in the United States over the past 20 years. During the early 1970s, DARPA sponsored a research program in connected-speech understanding capability, the Speech Understanding Research (SUR) program. DARPA specified general goals for the developers—the systems were to be able to accept continuous speech in a 1,000-word vocabulary relating to a particular task or domain area and were to be able to respond quickly with less than 10 percent error. The HEARSAY system, a product of this research effort, met these goals and pioneered the blackboard technique of organizing knowledge sources within systems.

DARPA's Image Understanding Program, which was launched in 1975, sponsored research into a theory of vision and into hardware for processing visual images. The intention of the Image Understanding Project was to integrate image processing, pattern recognition, computer science, artificial intelligence, neurophysiology, and physics in the attempt to find methods to automatically extract information from imagery. ACRONYM, a knowledge-based system for photo interpretation, was one product of the Image Understanding Program.

Currently, DARPA is sponsoring artificial intelligence research within a framework of specific defense applications in the Strategic Computing Program. The program aims to create a new generation of extremely powerful machine intelligence technology to support the United States' military security and ability to compete in the world economy. The research will be carried out by contract with government, industry, and university laboratories. The results of this research will be communicated to private industry for eventual exploitation and military applications. The program is expected to cost approximately $600 million over the first 5 years. It is estimated that in over 10 years, the cost will be $1 billion.

The first projects, which DARPA is contracting out to university laboratories, are in the areas of vision, speech, natural language, expert systems technology, and robotics. Industry and academic participants will cooperate in joint projects to develop the advanced computer architectures that will be necessary to carry out these programs.

- *The National Bureau of Standards*

The National Bureau of Standards is involved in research on interface standards in automated manufacturing. The bureau is also building a testbed for a flexible manufacturing system, the Automated Manufacturing Research Facility Project (AMRF). The goal of the program is to support manufacturing systems research by academia, industry, and other agencies; to conduct continuing studies of interface standards; and to transfer technology to American industry.

Major European Programs

- *ESPRIT*

The European Economic Community (EEC) Common Market countries formed the European Strategic Program for Research in Information Technology (ESPRIT) in 1984. ESPRIT is a collaborative of industry, universities, and governments reaching across national borders, and now including three American firms, ITT Corporation, Digital Equipment Corporation, and IBM Corporation in particular projects.[1]

The goal of ESPRIT is to promote European industrial cooperation in research and development in information technologies, and to attempt to make European standards prevail when the fifth-generation machine arrives on the scene.

Total funding for the first 5 years of the project, 1984 through 1988, is about $1.5 billion. Half of the funds comes from the EEC, the remainder from participating companies. The money is being distributed to research labs and universities for projects in the areas of microelectronics, robotics, software, information processing, office automation, and computer-aided manufacturing. Each project is supposed to be a collaboration between industrial partners from at least two countries plus one university.

- *A Joint European Project*

The three largest European mainframe computer manufacturers, ICL of Britain, Bull of France, and Siemens of West Germany, are cooperating on research in implementing advanced knowledge-processing systems. These systems could be used in decision-support systems for managers. The participating companies will have equal access to results of the research, but commercial applications will be developed and marketed by the companies individually.

Major United Kingdom Programs

- *The Alvey Program of Advanced Information Technology*

The Alvey Program, named for its chairman, John Alvey, is the United Kingdom's fifth-generation project. The goal is to develop the necessary technologies, emphasizing knowledge-based systems, in 5 years in contrast to Japan's 10. The Alvey Program, begun in 1982 after the Japanese announced their Fifth-Generation Computer Systems Project, aims to mobilize the United Kingdom's technical strengths in information technology in order to improve the country's competitive position in world information technology markets. The program coordinates efforts by industry, government bodies, universities, and research laboratories.

The Alvey program supplements but does not duplicate the United Kingdom's role in ESPRIT. The aims of the program are to establish the tools and methods necessary for the production of high-quality, cost effective hardware and software. The research areas are software engineering, VLSI, expert systems and intelligent, knowledge-based systems, and man-machine interfaces.

The program is managed by the Alvey Directorate whose staff comes from United Kingdom industry and from government funding bodies. The government and industry together are estimated to be providing a total budget for the first 5 years of about $500 million.

The directorate awards research monies to applicants bidding for the contracts. The applicants are expected to form research consortia, carrying out the research at their own facility, rather than at a central facility. An example is a consortium that is producing new integrated software engineering tools. It is composed of six industrial and academic members under the leadership of SDL, a software house.

One hardware project being carried out under the Alvey program is the creation of a microprocessor chip, called the "transputer," designed for parallel-processing applications.

- *The Turing Institute*

The Turing Institute at the University of Strathclyde in Edinburgh, Scotland, named after the British mathematician Alan Turing, is a research and teaching organization established in 1983.

The goals of the Turing Institute, under the leadership of director Donald Michie (also a professor at the University of Edinburgh), are to advance the education and knowledge of the public in addition to advancing industrial and technological concerns through research, development, teaching, and other scientific work in machine intelligence and computer technology. The training programs of the Turing Institute include Inference Systems and Logic Programming, Computer Vision, Expert Systems, and Robot Planning. The Institute also has a software library and computing facilities for members.

The Institute's members will also carry out research in the areas of expert systems, machine learning, robotics, and computer vision.

West Germany

West Germany's Ministry for Research and Technology has designed a program of research that brings industry and academia together in projects. The government's budget for the 5 years, 1984 through 1988, is about $1 billion. Research areas include computer-aided design, new computer structures, and knowledge engineering.[2]

Hungary

Although the Prolog language was invented in Edinburgh and Marseilles, the first practical Prolog applications were developed in Hungary. These applications are considered the first European work in expert systems development.

Hungary's Institute for Computer Coordination (SZKI) in Budapest has been the source of many programs in Prolog, beginning in 1977. Since then, SZKI has directed the development of over 250 expert systems written in Prolog for applications such as information retrieval in many areas—architectural design, pharmaceutical design and drug interactions, biochemical testing and modeling, COBOL program generation, program analysis, Prolog program verification, the parsing of texts in the Hungarian language, mathematics, and managing collective farms. The experience gives Hungary an advantage in the fifth-generation race, but the lack of hardware is a damper on the program.

In 1979, SZKI implemented a new version of Prolog, MPROLOG (Modular Prolog). The Japanese are using this version in their fifth-generation program. MPROLOG is distributed internationally by Logicware, Toronto, Ontario.

Though Hungary is not part of the European Economic Community, the eastern bloc nation has asked to participate in various ESPRIT ventures. Hungary's Prolog experience would be an asset to ESPRIT. However, political considerations may preclude a cooperative association.

Hungary's experience with Prolog and expert systems is also the reason Hungary is an asset to the Soviet Union in its attempts to enter the fifth-generation race.[3]

The Soviet Union

The USSR is organizing research among countries that make up CMEA (the Council for Mutual Economic Assistance). Input to the Soviet plan has come from Bulgaria, East Germany, Hungary, Poland, Czechoslovakia, Cuba, and Rumania.

The Soviet Union's 5-year plan to develop its own fifth-generation computer project is said to have a much lower budget than those in Japan, the United States, and the United Kingdom. The Commission for Computer Engineering at the Moscow Academy of Sciences has set goals for the plan in the areas of VLSI microprocessors, parallel and multiprocessor computer architectures, intelligent databases and methods of operation, and logic as the programming basis for computer operation.[4]

Annotated Bibliography

Barr, Avron, and Edward A. Feigenbaum (eds.), *The Handbook of Artificial Intelligence,* Volume III, Stanford, California, HeurisTech Press, 1982, pp. 135-136. The overview to the chapter on vision discusses the DARPA program on Image Understanding.

Bernhard, Robert, "The Fifth Generation – Awesome Obstacles," *Systems & Software,* January 1985, pp. 132-138. This article discusses hardware advances that will be necessary in order to make fifth-generation computers possible.

Feigenbaum, Edward A., and Pamela McCorduck, *The Fifth Generation,* Reading, Massachusetts, Addison-Wesley Publishing Co., 1983. This book discusses efforts under way around the world to develop a fifth-generation computer and associated advanced computing technologies. It focuses in particular on the need for the United States to compete with the Japanese program.

Highberger, Deb, and Dan Edson, "Intelligent Computing Era Takes Off," *Computer Design,* September 1984, pp. 79-95. This article surveys the efforts of the national, international, and intercorporational efforts to develop a fifth-generation computer and advanced supercomputer technologies. It also discusses the effects that the achievement of these goals would have on national defense, economies, and lifestyles.

Jones, Keith, "Knowledge Engineering: AI with a European Flavor," *Electronic Business,* November 1983. This short article describes a European AI collaborative composed of a British, a West German, and a French company.

Kawanobe, K., "Present Status of the Fifth Generation Computer Systems Project," *ICOT Symposium – Report,* based on lectures given at the ICOT symposium held June 21, 1984, pp. 13-21. The article outlines the current status of research and development in the Japanese Fifth-Generation Computer Systems Project.

Lamb, John, "Alvey Is on Its Way," *Datamation,* June 15, 1984, pp. 60, 84. This article discusses the launching of the Alvey program and reactions to it.

Santane-Toth, E. "PROLOG Applications in Hungary," In K.L. Clark and S.A. Tarnlund (eds.), *A.P.I.C. Studies in Data Processing, No. 16, Logic Programming.* 1982, Academic Press, New York. This paper provides an overview of Prolog applications in Hungary, describing the problems addressed, the main features of the implementations, and the results.

Shimoda, H., "Fifth-Generation Computer: From Dream to Reality," *Electronic Business,* November 1, 1984, pp. 68-72. This article presents an interview with Kazuhiro Fuchi, research director of the Institute for New Generation Computer Technology (ICOT). In the interview, Fuchi discusses the results of ICOT's work through the first 2 years of the fifth-generation project.

Tate, Paul, "The Blossoming of European AI," *Datamation,* International Edition, November 1, 1984, pp. 85-88. This article discusses AI research efforts in Europe, in industry, universities, and government and cooperative programs.

"The Race to Build a Supercomputer," *Newsweek,* July 4, 1983, pp. 58-64. This article tells how Japan and the United States are rushing to produce a new generation of supercomputers and computers designed for artificial intelligence work. It includes descriptions of US and Japanese programs in electronics and computing and descriptions of the new computer architectures.

Uttal, Bro, "Here Comes Computer, Inc.," *Fortune,* October 4, 1982, pp. 82-91. This article discusses the Japanese fifth-generation computer project: origins, leadership, and participants.

Walton, Paul, "Hungary Stretches a Hand Out to the West," *Computing,* May 24, 1984, p. 27. This article describes Hungary's experience in developing practical expert systems, in implementing and selling MPROLOG, and in efforts to join the fifth-generation computer race.

Walton, Paul, "Piggy in the Middle of a Power Struggle," *Computing* (U.K.), May 24, 1984, p. 26. This article discusses the Soviet Union's fifth-generation computer project and the potential for the United Kingdom to act as a technology broker between the East and the West.

Walton, Paul, and Paul Tate, "Soviets Aim for 5th Gen," *Datamation,* July 1, 1984, pp. 53-61. This article discusses the collaboration of the Soviet Union and other Eastern-bloc countries in a fifth-generation computer program.

Withington, Frederic G., "Winners and Losers in the Fifth Generation," *Datamation,* December 1983, pp. 193-209. This article describes emerging technologies in large-scale integrated circuits, disk drives, supercomputers, and knowledge-based or problem-solving systems and the role of national governments in guiding and supporting computer research.

Notes

1. Braggar, Hans, "Major U.S. Firms Announce Participation in ESPRIT," *Computerworld*, November 12, 1984, p. 46.

2. Woolnough, Roger, "Disturbing Signs Surface at European Fifth-Generation Meeting," *Electronic Engineering Times*, October 22, 1984, p. 22.

3. Walton, Paul, "Hungary Stretches a Hand Out to the West," *Computing*, May 24, 1984, p. 27.

4. Walton, Paul, "USSR Makes AI Plans," *Computing*, (UK), May 17, 1984, p. 3.

Appendix

Highlights in the Development of Artificial Intelligence

The following list, necessarily limited and selective, is a chronology of significant events in the brief history of AI. Sources of the information cited below include books in the annotated bibliography, communication with individuals in the institutions cited, and books and papers cited within this chronology. An attempt has been made to verify information by reference to either multiple or original sources. We regret any inadvertent errors.

1950
Alan Turing presented a scientific paper on the subject of artificial intelligence, *Computing Machinery and Intelligence*. In this paper, he proposed his test ("Turing's Test") for determining whether a machine possesses artificial intelligence. In an earlier paper, Turing had suggested that the brain could be simulated.

1955
IPL-II (Information Processing Language-II), the first AI language, was created by Allen Newell, J.C. Shaw, and Herbert Simon. IPL is a list-processing language.

1956
The Dartmouth Summer Conference on Artificial Intelligence, organized by John McCarthy, Marvin Minsky, Nathaniel Rochester, and Claude Shannon, with funds from the Rockefeller Foundation, brought together people whose work founded the field of AI. Among the participants in addition to the four organizers were Arthur Samuel, Trenchard More, Oliver Selfridge, Allen Newell, Ray Solomonoff, and Herbert Simon.

The Logic Theorist (LT), developed by Newell, Shaw, and Simon, was discussed at this conference. The LT, considered the first AI program, used heuristic search to solve problems in Whitehead and Russell's *Principia Mathematica*.

Mid–late 1950s
John McCarthy, then at MIT, designed the LISP (LIST PROCESSING) language.

1957
Newell, Shaw, and Simon began developing the General Problem Solver (GPS). This program applied codified problem-solving techniques, including means-ends analysis, to problems in a number of different task environments.

1959
Culminating years of experimentation, Arthur Samuel completed a checkers-playing computer program that performed as well as some of the highest-rated players of that time. His paper, entitled "Some Studies in Machine Learning Using the Game of Checkers" was published in the *IBM Journal of Research and Development*.

1959
Frank Rosenblatt's paper describing his pattern-recognition machine, the Perceptron, was published in *Proceedings of a Symposium on the Mechanization of Thought Processes*. The paper was entitled "Two Theorems of Statistical Separability in the Perceptron."

1960
Research began in the MIT Artificial Intelligence Project under the direction of John McCarthy and Marvin Minsky.

1963
Computers and Thought, Edward. A. Feigenbaum and Julian Feldman, (eds.), was published. Marvin Minsky's article, "Steps toward Artificial Intelligence," was included in this collection.

1964
Daniel G. Bobrow published his Ph.D. thesis, based on his system STUDENT. STUDENT is a natural language program that can understand and solve high school algebra story problems.

1965
The Stanford University Heuristic Programming Project (HPP), an artificial intelligence research laboratory within Stanford's Computer Science Department, began research in expert systems. The HPP is

now part of Stanford's Knowledge Systems Laboratory. Edward A. Feigenbaum is currently the principal investigator in the HPP.

1965
Work began on DENDRAL, the first expert system. Developed at Stanford University by a group including Joshua Lederberg, Edward A. Feigenbaum, Bruce G. Buchanan, Dennis Smith, and Carl Djerassi, DENDRAL analyzes information about chemical compounds to determine their structures.

1966
"ELIZA – A Computer Program for the Study of Natural Language Communication between Man and Machine" was published in *Communications of the Association for Computing Machines*. Joseph Weizenbaum created ELIZA to illustrate that natural-language capabilities can make a computer seem deceptively intelligent. ELIZA was a psychology program that simulated the responses of a therapist in interactive dialogue with a "patient."

1966
Richard D. Greenblatt began developing a computer chess game capable of competing successfully in tournaments. This system was described in "The Greenblatt Chess Program," in *AFIPS Conference Proceedings*.

1966-1972
SHAKEY, a mobile robot, was built at SRI International. Shakey's "intelligence" allowed it to perceive and to plan actions in order to carry out tasks.

1968
Marvin Minsky's *Semantic Information Processing* was published. One of the programs described in the book, developed by Minsky's student Thomas G. Evans, could answer geometry analogy questions from an IQ test. "Semantic Memory" by M. Ross Quillian, which discussed his semantic network concept, was also included in the volume. Quillian used semantic nets to model human associative memory.

1970
Patrick H. Winston's Ph.D. thesis, *Learning Structural Descriptions from Examples,* was published. The thesis describes ARCHES, a program that learned from examples.

1970

MIT's Artificial Intelligence Project became MIT's Artificial Intelligence Laboratory under the direction of Marvin Minsky and Seymour Papert. The laboratory has been under the direction of Patrick H. Winston since 1973. Earlier work on artificial intelligence at MIT led to the first basic tools for word processing and the concept of time-sharing computers. Current research at MIT includes computer vision, all areas of robotics, expert systems, learning and common-sense reasoning, natural language, and computer architectures.

1970

Jack D. Myers and Harry E. Pople began work at the University of Pittsburgh on INTERNIST, now called CADUCEUS, a system intended to aid physicians in the diagnosis of human diseases.

1970

Alain Colmerauer and his colleagues began developing the Prolog programming language. Prolog development has also been active in Edinburgh, London, and Budapest.

1970

Terry Winograd, then at MIT, wrote for his Ph.D. thesis SHRDLU, a natural language-understanding program that could respond to questions and plan actions in a simplified "blocks world." The thesis was later published as *Understanding Natural Language*.

1971

Nils Nilsson and Richard Fikes completed work on STRIPS at SRI International. STRIPS made use of plans, sequences of operators, to achieve goals.

1971

MACSYMA was first used. MACSYMA was developed over more than a decade at MIT by William Martin and Joel Moses. MACSYMA's design was based on prior work by Martin, Moses, and Carl Engleman. MACSYMA performs differential and integral calculus and simplifies symbolic expressions. Both inputs and outputs are symbolic and the program is knowledge-based. This program is widely used by mathematicians, research physicists, and engineers.

1971-1976
The United States Defense Advanced Research Projects Agency (DARPA) sponsored research into connected-speech understanding capability in the Speech Understanding Research (SUR) Program. Some of the resulting programs were SPEECHLIS and HWIM (Hear What I Mean), from Bolt Beranek and Newman, Inc., and HEARSAY-I, HEARSAY-II, DRAGON, and HARPY from Carnegie-Mellon University.

1972
William Woods et al., at Bolt Beranek and Newman, developed LUNAR, an information retrieval system that uses augmented transition networks (ATNs) as the representation form for its natural language system grammar. LUNAR was intended for use by geologists in the evaluation of materials obtained from the moon during the *Apollo-11* mission. Woods had developed the ATN concept earlier in a 1970 paper on the subject.

1973
SUMEX-AIM (Stanford University Medical Experimental Computer Project—Artificial Intelligence in Medicine) was formed as a community resource for the development of AI techniques with support from the National Institutes of Health. SUMEX-AIM has been the source of MOLGEN and other projects in medicine, biochemistry, and psychology.

1973
Roger C. Schank's "Conceptual Dependency: A Theory of Natural Language Understanding" was published in *Cognitive Psychology*. Schank et al., at the Stanford University AI Laboratory, later used the conceptual dependency knowledge representation in MARGIE, a natural language understanding program that could make inferences and generate paraphrases.

1973
Computer Models of Thought and Language, Roger C. Schank and Kenneth M. Colby (eds.), was published.

1973
Pattern Classification and Scene Analysis, by R.O. Duda and P.E. Hart was published.

mid-1970s

The initial version of MYCIN, an expert system that makes recommendations for the treatment of meningitis and other bacterial infections in the blood, was developed within SUMEX-AIM by Edward H. Shortliffe. MYCIN's medical knowledge is encoded as production rules.

1975

DARPA began the Image Understanding Program to sponsor research into machine vision, including developing a theory of vision and hardware for image processing. ACRONYM, a model-driven interpretation system, was developed under this program by R.A. Brooks.

1975

The Psychology of Computer Vision, (Patrick Winston, ed.) was published. Marvin Minsky's paper, "A Framework for Representing Knowledge," was included in this collection. Minsky's paper discussed frames as useful structures for organizing knowledge in many systems including natural language and vision systems. The collection also included "Understanding Line Drawings of Scenes with Shadows," by David Waltz. This paper discussed a new way to use the edges of shadows to interpret visual images.

1975

Roger C. Schank and Robert Abelson et al., at Yale, published a paper describing SAM (Script Applier Mechanism), a natural language understanding program that added the use of scripts to conceptual dependency representations.

1975

Representation and Understanding, Daniel G. Bobrow and Allan Collins (eds.), was published. This volume included important papers on knowledge representation.

1976

Douglas B. Lenat wrote AM, a type of learning program that defines and evaluates mathematical concepts in set and number theory. This process has been described as "automated discovery."

1976

Randall Davis published his thesis for a Ph.D. at Stanford University on TEIRESIAS, a system that utilizes metalevel knowledge to enter and update knowledge bases used in expert systems. The thesis, published as a Stanford AI Memo, was later published in *Knowledge-Based Systems in Artificial Intelligence,* Randall Davis and Douglas B. Lenat, joint authors.

1977
Programmers at Hungary's Institute for Computer Coordination (SZKI), in Budapest, completed the first of many practical expert system applications utilizing the Prolog language.

1978
R.O. Duda et al., at Stanford Research Institute International, published a paper discussing PROSPECTOR, an expert system that assists in the analysis of information related to geological exploration.

1980
XCON, the first expert system successfully used on a daily basis in a commercial environment, went into operation at Digital Equipment Corporation. The prototype for XCON was developed under the direction of John McDermott of Carnegie-Mellon University.

1981
The first volume of the three-volume set, *The Handbook of Artificial Intelligence,* Avron Barr (ed.) Vols. I and II, Edward A. Feigenbaum (ed.) Vols. I-III, and Paul R. Cohen (ed.) Vol. III was published. The other two volumes of the handbook were published the following year.

1981
Japan announced its intention to organize a Fifth-Generation Computer Systems Project.

1982
Japan's Institute for New Generation Computing Technology (ICOT) was formally launched at its Tokyo headquarters.

1982
The Microelectronics and Computer Technology Corporation (MCC) was formed in the United States to respond to the Japanese fifth-generation program.

1982
The United Kingdom began the Alvey Program of Advanced Information Technology to perform fifth-generation computer research.

1983
The European Economic Community formed ESPRIT to compete in the race to develop a fifth-generation computer.

1983
MCC opened for business in Austin, Texas.

1983
The Turing Institute opened at the University of Strathclyde in Edinburgh, Scotland, offering training in subjects related to machine intelligence.

Annotated Bibliography

For more information on the history of artificial intelligence, see the following books:

Barr, Avron, and Edward A. Feigenbaum (eds.), Vols. I and II, Paul R. Cohen and E. A. Feigenbaum (eds.), Vol. III, *The Handbook of Artificial Intelligence*, Stanford, California, HeurisTech Press, 1981, (C) by William Kaufman, Inc., Los Altos, California, Vol. I, 1981, and Vols. II and III, 1982. This three-volume work surveys the field of AI research, discussing concepts, techniques, and the importance of many groundbreaking programs.

Boden, Margaret A., *Artificial Intelligence and Natural Man,* New York, Basic Books, Inc., 1977. This book discusses concepts of AI in the context of illustrative programs that do interesting things, mostly in psychology or philosophy. Boden stresses that AI may help us solve problems in the philosophy of mind and illuminate the complexities of human psychology.

McCorduck, Pamela, *Machines Who Think,* W. H. Freeman and Company, San Francisco, 1979. This is a very readable introduction to the history of artificial intelligence and automata.

Newell, Allen, and Herbert A. Simon, *Human Problem-Solving,* Prentice-Hall, Inc., Englewood Cliffs, New Jersey, 1972. This book explores issues of how humans think and sets forth a theory of human problem-solving. It includes an historical addendum.

Glossary

Algorithm. An explicit, finite set of instructions for solving a problem.

Artificial intelligence (AI). *A broad definition*: A growing set of computer problem-solving techniques that are being developed to imitate human thought or decision-making processes or to produce the same results as those processes.

Augmented Transition Network (ATN). A grammar representation used in natural language systems to parse input. ATNs are strings representing legal sequences of parts of speech. Input text is compared to the strings. If the parts of speech represented by words in the input match the parts of speech in an ATN string, the system recognizes the input as legal. Actions to be performed on or with input may be associated with an ATN.

Automatic program synthesis. The automatic computer construction of programs from rigorous and nonalgorithmic specifications describing what the programs should do.

Automatic program verification. The use, by a computer, of mathematical techniques to show that programs behave according to their formal specifications, in order to prove the correctness of programs.

Automatic programming. The application of artificial intelligence techniques to the general goals of automatic program construction and automatic program transformation.

Backtracking. Returning the database, or conditions, in a system to a previous state in order to try an alternative solution path.

Backward chaining. Attempting to solve a problem by stating a goal and looking in the database for the conditions that would cause it to come about, then reiterating this process, using those conditions as the goals and searching for their preconditions, and so on.

Blackboard. A data structure on which a system can post information on internal states of objects or registers in the system for consultation and appropriate action by operators in the system.

Certainty factor. A tag associated with a piece of information supplied to an expert system. The tag describes the application developer's degree of confidence that the piece of information is likely to be true.

Conceptual dependency. A knowledge representation used in natural language systems. A conceptual dependency translates input into an internal representation usable by the system. A conceptual dependency makes use of a small number of basic components signifying meaning that can be combined to represent more complex meanings.

Confidence factor. *See* Certainty factor.

Connected-speech system. A system that can understand a stream of speech in which the speaker does not pause between words to emphasize the beginnings and ends of the words.

Deduction. The process of reaching a conclusion by logical means.

Document generation system. A system that synthesizes text by manipulating stored information in response to specifications for the output. A type of document understanding system.

Document understanding system. A system that accepts textual input, stores and classifies it, and then utilizes the input in response to specifications for tasks. Some document understanding systems provide paraphrases, some answer questions, some perform translation, and some draw inferences.

Domain. The problem area about which a system has knowledge and an ability to manipulate the knowledge.

Ellipsis. The omission of words so that a sentence is not grammatically complete but is still comprehensible to the intended audience.

Expert system. A computer system that solves problems by manipulating knowledge encoded from expert sources.

Explanation facility. A feature of many expert systems. The part of a system that tells what steps were involved in the processing by which it arrived at a solution. These facilities can be simple traces of steps or they can be more complex, supplying encoded reasons and references for the reason the solution takes one alternative rather than another.

Fifth-generation computer. An upcoming generation of computers that will not depend on a von Neumann architecture. The first four generations of computers processed instructions sequentially. Definitions of the fifth-generation computer vary, depending on the goals of the researchers working on different fifth-generation projects, but most include the ability to do parallel processing and an architecture suited to knowledge processing.

Forward chaining. The type of activity done in a system that applies operators to a current state in order to produce a new state, and so on until the solution is reached.

Frame. A knowledge representation based on the idea of a frame of reference. A frame carries with it a set of slots which can represent objects that are normally associated with the subject of the frame. The slots can then point to other slots or frames. This gives frame systems the ability to carry out inheritance and simple kinds of inferencing.

Goal. The solution a system attempts to reach using operators. Sometimes, in order to reach the goal, subgoals must first be achieved.

Heuristic. (Used as a noun in AI.) A process, sometimes a rule of thumb, that may help in the solution of a problem, but that does not guarantee the best solution, or indeed, any solution.

Icon. A symbol to which a computer user can point an interface device in order to select a function, such as "move window."

Image processing. The examination by a computer of digitized data about a scene and the features in it in order to extract information.

Inference. A conclusion based on a premise.

Inference engine. The part of a rule-based system that selects and executes rules.

Intelligent system. A system equipped with a knowledge base that can be manipulated in order to make inferences.

Isolated-word system. A speech-understanding system whose input must consist of words enunciated separately, instead of run together, to make word identification easier for the system.

Knowledge base. The part of a knowledge-based system that contains codified knowledge and heurstics used to solve problems.

Knowledge engineering. The activities of software engineers who acquire knowledge for knowledge-based systems and decide how to represent it for use in the system.

Knowledge representation. A structure in which knowledge can be stored in a way that allows the system to understand the relationships among pieces of knowledge and to manipulate those relationships.

Machine translation. The translation by computer of text in one human native language into text in another language.

Metaknowledge. Knowledge about knowledge. Knowledge that tells a system something about what it knows, how its knowledge can be utilized, and what the limits of its knowledge are.

Natural language. A person's native tongue. Natural language systems attempt to make computers capable of processing language the way people normally speak it instead of requiring specialized programming languages.

Object-oriented programming. Programming that focuses on objects rather than on procedures.

Parallel processing. Computer processing that carries out more than one instruction at a time, rather than carrying them out sequentially.

Parsing. Identifying the components of language statements as various parts of speech.

Predicate logic. A form of logic that can be utilized in knowledge-based programs. The basis of the AI language Prolog.

Production rule. A condition/action rule that produces change in a system that results in new conditions and new actions, and so on. Most rules are structured in an "If...then...." format.

Protocol. A set of rules for accomplishing an activity.

Robot. A mulitifunctional manipulator that can be programmed and reprogrammed to perform tasks.

Rule. *See* Production rule.

Script. A knowledge representation form based on stereotyped situations. When activities fit into stereotypes, a system equipped with a script can predict other likely events by analogy, "assuming" that the script will continue to hold true.

Search. The process of trying different actions in a system until a sequence of actions is discovered that will achieve a goal state.

Semantic net. A knowledge representation composed of nodes, describing objects, and links, describing relationships between nodes.

Semantics. The meaning of words within context.

Sensing system. A system that processes digitized signal input about shape, weight, pressure, friction, temperature, and other sensation information. Sensing systems interpret the input by comparing it to stored patterns.

Slot. *See* Frame.

Speaker-dependent system. A speech-generation system that is trained to recognize words as they are spoken by specific individuals on the basis of training with samples of that speaker's pronunciation.

Speech-generation system. A system that translates text into audible speech with correct pronunciation.

Speech-understanding system. A system that converts the digitized signals of audible speech into printed text.

Syntax. The structure into which words fit.

Vision system. A system that interprets digitized input about the shape, location, and sometimes the color of objects, in order to determine what the input represents and/or what the significant features of the object are.

Working memory. In OPS5, the dynamic portion of a production system's memory. Working memory contains the changing database of the system as rules fire.

Index

A* algorithm search procedure, 62
Abelson, Robert, 166
abstraction methods, 68
"active sensing," 37
ACRONYM (vision system), 166
address space, 97
Advanced Research Projects Agency (ARPA; U.S. Department of Defense), 30, 81, 152-53, 165
"advanced systems," 8
AGE (software tool), 138
AI Spear (Computer System Failure Analysis Tool; expert system), 123
AL (language), 79
algorithms, 9
  in automatic programming, 42
  in development of expert systems, 127, 128
ALPS (machine translation) system, 28
Alvey, John, 154
Alvey Program of Advanced Information Technology (United Kingdom), 154-55, 167
AM (learning program), 166
AML (language), 79
application development
  hardware for, 98-105
  software tools for, 135-40
applications, 15
  automatic programming and intelligent programming aids, 40-45
  expert systems, 15-21, 111
  languages for, 77
  natural language systems, 21-32
  of object-oriented programming, 92-93
  robotics and sensing systems, 32-40
  see also expert systems
arcs
  in augmented transition networks, 70
  in semantic networks, 66
ARCHES (learning program), 163

ART (Automated Reasoning Tool), 137
artificial intelligence
  chronology of, 161-68
  computer hardware for, 97-108
  definitions of, 7-9
  expert system applications of, 15-21
  languages for, 73-74, 77-93
  national programs and cooperative research in, 147-57
  problem-solving techniques for, 51-53
  techniques and devices for, 68-73
  training in, from Digital Equipment Corporation Educational Services, 141-43
Artificial Intelligence Corporation, 25
atoms, in LISP, 80
augmented transition networks (ATNs), 70
Automated Language Processing Systems, Inc., 28
Automated Manufacturing Research Facility Project (AMRF; National Bureau of Standards), 35, 153
automatically guided vehicles (AGVs), 34
automatic code generation, 41
automatic programming, 40-45
Automatix, Inc., 33, 79

backward chaining, 19, 70-71
Baroid, NL, 9-10
Barr, Avron, 167
BASIC, Machine Intelligence, 79
beam searches, 62
Belle (chess-playing system), 10
Bell Laboratories (AT&T), 10, 106
best-first searches, 59
Binford, Thomas, 79
BIS (bidirectional synthesizer), 45
blackboards, 72
BLISS (language), 89
Bobrow, Daniel G., 78, 162, 166

Bolt Beranek and Newman, Inc., 82
BORIS (document understanding system), 29
bottom-up processing (forward chaining), 70
branch-and-bound searches, 62
breadth-first searches, 59, 61
Brooks, R.A., 166
Buchanan, Bruce G., 8, 163

CADUCEUS (medical expert system), 164
Carnegie Group, Inc., 137, 139
Carnegie-Mellon University, 10, 19, 34, 139
Caruso, Richard, 113, 115, 116
chaining, 70-71
chess-playing systems, 10, 57
CHI (automatic programming system), 43-44
classes, in object-oriented languages, 90
clauses, in Prolog, 84-85
CLISP (Common LISP; language), 83
CLU (language), 92
Cognitive Systems, Inc., 25-26
Cohen, Paul R., 167
Colby, Kenneth M., 165
Collins, Allan, 166
Colmerauer, Alain, 83, 164
Colon Osorio, Fernando C., 143
combinatorial explosion, 57-58
common language interfaces, 100
COMMON LISP (language), 81-82
"common sense," 11
communications
  blackboards for, 72
  DECnet for, 107
  using natural language, 21-22
compatibility, of hardware, 99
compilers
  for LISP, 81
  optimizing, 41
computers
  for application development, 98-102
  for artificial intelligence applications, 97-98
  from Digital Equipment Corporation, 105-8
  document understanding by, 28-29
  fifth-generation, 148-50, 154-55, 157
  natural language communications with, 21-22
  natural language translation by, 27-28
  for research and development, 102-5
  XCON expert system for configuration of, 19-20, 112-13
conceptual dependencies, 28, 66-68
Concurrent Prolog (language), 149
confidence factors, 72-73
configuration of computers, XCON expert system for, 19-20, 112-13
conflict resolution, in OPS5, 88
CONNIVER (language), 78
constants, in logic, 62
consulting services, from Digital Equipment Corporation, 141
control strategies, 71-72
C-Prolog (language), 87
CYRUS (document understanding system), 29

Dartmouth Summer Conference on Artificial Intelligence, 161
data, 53
  in procedure-oriented versus object-oriented programming, 90
databases, 52
  in OPS5, 88
  in Prolog, 85-86
Davis, Randall, 166, 167
DEC-10 Prolog (language), 87
DEC-10/20 Prolog (Prolog-20; language), 84, 87
DECnet, 107
DECOR (graphics language), 107
DECsystem-10 computers, 115
DECsystem-10/DECSYSTEM-20 Prolog (language), 84, 87
DECSYSTEM-20 Common LISP (CLISP; language), 83
DECSYSTEM-20 computers, 107-8, 115
DECtalk (speech generation system), 32
deduction-driven synthesis based on transformation rules, 42
default values, 65
Defense Advanced Research Projects Agency (DARPA; U.S. Department of Defense), 30, 81, 152-53, 165
DeJong, Gerald, 29
delivery, of expert systems, 134-35
delivery systems, 102
demons, 72
DENDRAL (expert system), 163

depth-first searches, 59, 61
development and maintenance teams, for expert systems, 130-31
dialects
 of LISP, 82-83, 136
 in LISP machines, 105
 object-oriented, 91
 of Prolog, 87
dictionaries
 in machine translation systems, 27-28
 of speech generation systems, 32
 in text critiquing systems, 30
Digital Equipment Corporation
 artificial intelligence applications on computers by, 105-8
 DECtalk by, 32
 Educational Services of, 139
 OPS5 from, 89-90
 Prolog for computers by, 87
 resources for development of expert systems from, 140-44
 VAX LISP and GCLISP from, 82
 XCON expert system by, 19-20, 112-26
Digital Press, 143-44
documents
 critiquing of, 30
 machine translation of, 27-28
 machine understanding of, 28-29
Djerassi, Carl, 163
DUCK (software tool), 138
Duda, R.O., 165, 167

Educational Services (Digital Equipment Corporation), 141-43
ellipsis, in natural language, 24
ELIZA, 163
EMYCIN (software tool), 138
Engleman, Carl, 164
environments, programming, 135
EPISTLE (text critiquing system), 30
European Economic Community (EEC; Common Market), 153-54
European research programs, 153-54
European Strategic Program for Research in Information Technology (ESPRIT), 153-54, 168
EUROTRA (machine translation system), 28
Evans, Thomas, 163
EXPERT-EASE (software tool), 138
experts (human), 126, 127, 130

knowledge engineers and, 132
expert systems, 12, 15-21, 111
 confidence factors in, 72-73
 development of, 53, 126-35
 explanation facilities in, 71
 MUDMAN, 9-10
 OPS5 for, 88
 production systems in, 56
 resources to aid in development of, 135-44
 XCON, 112-26
explanation facilities, in expert systems, 16-17, 71

factories, automated, 35, 153
feedback, in development of expert systems, 53, 133-34
Feigenbaum, Edward A., 8, 163, 167
Feldman, Julian, 162
fifth-generation computers, 148
 Alvey Program for, 154-55
 in Japan, 149-50, 167
 in Soviet Union, 157
Fikes, Richard, 35, 164
Flavors package (object-oriented extension to ZETALISP), 91
flexible manufacturing environments, 35
Forgy, Charles L., 88, 116
FORTRAN (language), 74, 79
forward chaining, 19, 70-71
 in OPS5, 89
Fox, Mark S., 139
frames, 18, 64-65
Franz LISP (language), 83
FRUMP (document understanding system), 29
Fuchi, Kazuhiro, 149, 150
functions
 in LISP, 80
 in logic, 62-63

game-playing systems, 10, 57-58
GCLISP (Golden Common LISP; language), 82, 108, 136
General Problem Solver, 162
general purpose workstations, 101
General Research Corporation, 138
GIST (automatic programming system), 44
GKS (graphics language), 107

178   Index

Glorioso, Robert M., 143
goals, 57
  in Prolog, 86
Gold Hill Computers, Inc., 82, 136
grammars, 69-70
  in natural language, 23
graphics
  in LISP machines, 104
  in VAX systems, 107
graphs
  of searches, 60-61
  of state-space, 57
Green, Cordell, 43
Greenblatt, Richard D., 163

HAM-ANS (Hamburg Application oriented Natural Language System), 26
Hamburg University, 26
hardware
  for application development, 98-102
  for artificial intelligence applications, 97-98
  from Digital Equipment Corporation, 105-8
  expert systems for testing, 123, 124
  for natural language communications, 22
  for research and development, 102-5
  XCON expert system for configuration of, 112-13
Hart, P.E., 165
HEARSAY-I (speech recognition system), 30
HEARSAY-II (speech recognition system), 30
HEARSAY III (software tool), 138-39
heuristics, 18, 58
  in vision systems, 36
Higher Order Software, Inc., 43
"high-level" vision systems, 36
hill-climbing searches, 59
human resources required in development of expert systems, 129-31
Hungary, 156

IBUS (Intelligent BUsiness System; expert system), 123-24
IDT (Intelligent Diagnostic Tool; expert system), 124

"If…then…." (production) rules, 18, 55-56, 88
Image Understanding Program, 152-53
industrial robots, 33-34
INET (Intelligent Network; expert system), 124
inference
  in definitions of artificial intelligence, 7-8
  rules of, 63, 64
Inference Corporation, 137
inference engines, 18-19, 88
information, 53
information management tools, 106
Information Sciences Institute (ISI; University of Southern California), 44, 136
inheritance, 68-69
  in object-oriented programming, 92
Inman, Bobby Ray, 151
Institute for New Generation Computing (ICOT; Japan), 149-50
INTELLECT (natural language interface), 25
IntelliCorp, 137
intelligent programming aids, 40-45
intelligent retrieval systems, 26
interfaces, natural language, 24-27
INTERLISP (language), 82-83, 104
INTERLISP-D (language), 91
INTERLISP for VAX (language), 136
International Business Machines Corporation (IBM), 30, 79
INTERNIST (medical expert system), 164
interpreters
  for LISP, 81
  for OPS5, 88
IPL-II (language), 78, 161
ISA (Intelligent Scheduling Assistant; expert system), 124

Japan, 148-50
Jeffrey Perrone & Associates, Inc., 138

KAS (Knowledge Acquisition System; software tool), 139
KBTA (Knowledge Based Test Assistant; expert system), 123
KEE (Knowledge Engineering Environment for Industry), 137
Kestrel Institute, 43

knowledge, 53
  encoding, for expert systems, 131-32
  representation of, 54-55
knowledge acquisition, by expert
  systems, 17-18
knowledge-based synthesis, 42
knowledge-based systems, *see* expert
  systems
knowledge bases, 18
  in document understanding
    systems, 29
  in OPS5, 88
  semantic information in, 23
Knowledge Craft (environment), 137
knowledge engineering tools, 135-39
knowledge engineers, 17-18, 130-32
knowledge representation, 18
Kolodner, Janet, 29
KRL (language), 78

Language Craft (environment), 137
languages (natural)
  document understanding in, 28-30
  Japanese, 149
  natural language systems, 21-32, 21-27
  speech generation of, 32
  speech understanding of, 30-32
languages (programming), 73-74, 77-79
  automatic programming and, 40
  as knowledge engineering tools,
    135-39
  LISP, 79-83
  object-oriented, 90-93
  OPS5, 19, 88-90
  production system, 56
  Prolog, 83-87
  for research and development systems,
    103-4
  supported on VAX systems, 106
language understanding, 22-23
Lederberg, Joshua, 163
Lehnert, Wendy G., 29
Lenat, Douglas, 166, 167
links, in semantic networks, 66
LISP (language), 43, 79-83, 162
  dialects of, 136
  object-oriented dialects of, 91
  OPS5 written in, 88
  Prolog compared with, 84
  training in, 140, 142
LISP Machine, Inc., 105, 140

LISP machines, 104-5
lists, in LISP, 80-81
logic (predicate calculus), 62-64
Logic Theorist (LT), 161
Logicware, 156
LOGOS (machine translation system),
  27-28
Logos Corporation, 27
LOOPS (environment), 91, 137-38
"low-level" vision systems, 36
LUNAR (natural language system), 165

McCarthy, John, 79, 161, 162
McDermott, John, 114, 167
  MUDMAN developed by, 10
  XCON developed by, 19, 89, 115, 116,
    118, 121, 125
Machine Intelligence BASIC (language), 79
Machine Intelligence Corporation, 79
machine translation of natural
  language, 27-28
MACLISP (language), 82, 104
MACSYMA (program), 82, 164
mainframe computers, 100-1
management, in development of expert
  systems, 130
Martin, William, 164
Marseilles, University of, 83
Massachusetts Institute of Technology
  (MIT), 44, 79, 82, 105
means-ends analysis, 62
medical applications of expert
  systems, 16
memory, 98
  in document understanding
    systems, 29
  in research and development
    systems, 103
  virtual, in VAX systems, 105
meningitis, expert system for treatment
  of, 16
messages, between procedures, 72
metaknowledge, 55
metaphors, 24
Michie, Donald, 155
Microelectronics and Computer
  Technology Corporation (MCC),
  150-52, 167, 168
MicroVAX II microprocessor, 107
Milan Polytechnic University, 45

180 Index

minicomputers, 101
Ministry of International Trade and Industry, Japan (MITI), 149, 150
Minsky, Marvin, 8, 64, 161, 162, 163, 166
mobile robots, 34-35
modus ponens, 63
Moses, Joel, 164
Moto-oka, Tohru, 149
MPROLOG (Modular Prolog; language), 87, 156
MUDMAN (expert system), 9-10, 89
Mumford, Enid, 129
MYCIN (expert system), 16, 166
Myers, Jack D., 164

National Bureau of Standards, U.S., 35, 153
natural language, 22
 Japanese, 149
 see also languages (natural)
natural language interfaces, 24-27
natural language systems, 21-27
 augmented transition networks in, 70
 conceptual dependencies in, 66-67
 document understanding in, 28-30
 machine translation in, 27-28
 Prolog used in, 84
 speech generation by, 32
 speech understanding in, 30-32
networking, 98, 102
 DECnet for, 107
 in research and development systems, 103
networks
 augmented transition, 70
 semantic, 66
Newell, Allen, 78, 161, 162
Nilsson, Nils, 8, 35-36, 164
NL Baroid, 9
nodes, 57
 in augmented transition networks, 70
 in semantic networks, 66
Norwegian Computing Center, 90
NTC (Network Troubleshooting Consultant; expert system), 124

object-oriented extensions, 91
object-oriented languages, 90-93
O'Connor, Dennis, 19, 113
Odetics, Inc., 34-35

operating systems, 106
OPS (language), 114
OPS4 (language), 114-16, 139
OPS5 (language), 19, 56, 73-74, 88-90, 139
 training in, 142
 for VAX systems, 136
 XCON in, 116
optimizing compilers, 41
output, natural language for, 21-22

Papert, Seymour, 163
parallel processors, 84
Parallel Prolog (language), 149
parsers, 69-70
parse trees, 69
parsing, 23, 24, 69-70
 in document understanding systems, 29
 in text critiquing systems, 30
PDP-6 mainframe computers, 100
PDP-11 systems, 113
Perceptron, 162
Pereira, Fernando C.N., 87
personal computers, 98, 102
 Rainbows, 108
Personal Sequential Inference machine (PSI), 150
perspectives (networks), 66
Pople, Harry E., 164
Prologia, 87
predicate calculus, 62-64
predicates, 63
primitives, 67-68
problem-reduction based synthesis, 42
problems
 analyzed by knowledge engineers, 132
 appropriate for expert systems, 126
 identification of, for expert systems, 131
 identification of, for XCON, 112-13
 in prototypes of expert systems, 133
procedural overloading, 92
procedure-oriented programming, 90
procedures
 control strategies for, 71-72
 in LISP, 80
 in object-oriented programming, 90
production ("If...then....") rules, 18, 55-56, 88
production memory, 88

production systems, 55-56
  in OPS5, 88-89
program development
  hardware for, 98-105
  LISP facilities for, 81
  software resources for, 135-39
programmers
  in development of expert systems, 130
  language preferences of, 77-78
Programmer's Apprentice (PA; software development system), 44-45
programming
  automatic, 40-45
  conventional, 51
  in OPS5, 56, 89
  procedure-oriented versus object-oriented, 90
  of robots, 35
programming environments, 135
programming languages, see languages (programming)
program synthesis systems, 41
Prolog (language), 64, 83-87, 156
Prolog II (language), 87, 108, 137
Prolog-20 (DEC-10/20 Prolog; language), 84, 87
Prologia, 137
pronunciation, by speech generation systems, 32
propositions, 62-63
PROSPECTOR (expert system), 129, 167
prototypes of expert systems, 53, 128, 132-33
PSL (Portable Standard LISP; language), 83
PUFF (expert system), 11

Quillian, M. Ross, 163
Quintus Prolog (language), 87, 137

R1 (XCON; expert system), 19
Raibert, Marc, 34
RAIL (language), 79
Rainbow personal computers, 108
Rand Corporation, 139
Reasoning Systems, Inc., 44
relational databases, 52
relationships
  between data, 52
  manipulation of, in artificial intelligence, 53

remote sensing, vision systems for, 36
research and development, computers for, 102-5
resources
  in development of expert systems, 135-40
  from Digital Equipment Corporation, for development of expert systems, 140-44
  human, in development of expert systems, 129-31
Rich, Charles, 44
robotics, 32-33
  future of, 147
  industrial, 33-34
  languages used in, 79
  mobile, 34-35
  teaching methods for, 35-36
  touch sensing, 39-40
  vision systems, 36-39
Robovision (robot), 33
Rosenblatt, Frank, 162
ROSIE (software tool), 139
rule-based systems, 55-56
rules
  of inference, 63, 64
  in OPS5, 88-89
  production ("If...then...."), 18, 55-56
  in Prolog, 85-86
  in XCON, sample of, 119

S.1 (software package), 138
SAIL (language), 78
SAM (natural language system), 166
Samuel, Arthur, 162
San Marco LISP Explorer (tutorial), 108
Schank, Roger C., 25, 28, 64, 67, 165, 166
scripts, 65-66
  in document understanding systems, 29
  in machine translation systems, 28
  in natural language interfaces, 25
searches, 58-62
  in Prolog, 87
semantic networks, 18, 66
semantics
  in natural language interfaces, 25
  of natural language, 23, 24
sensing systems, 32-33
  touch sensing, 39-40

sensing systems (cont.)
  vision systems, 36-39
SHAKEY (robot), 163
Shannon, Claude E., 57-58
Shapiro, Ehud, 149
Shaw, J.C., 78, 161, 162
shells, 135
Shortliffe, Edward, H., 166
SHRDLU (natural language system), 164
Simon, Herbert, 78, 161, 162
SIMULA 67 (language), 90
SLR (Schema Representation Language), 137
SMALLTALK-80 (language), 91
Smart Systems Technology, 138, 140
Smith, Dennis, 163
software
  availability of, 98
  knowledge engineering tools, 135-39
  operating systems, 106
software development systems, 44-45
Software Services (Digital Equipment Corporation), 140-41
Somalvico, Marco, 45
sound
  for natural language communication, 22
  speech generation, 32
  speech understanding, 30-32
Soviet Union, 157
Spear (Computer System Failure Analysis Tool; expert system), 123
speech generation, 22, 32
speech recognition and synthesis systems, 22
speech understanding, 22, 30-32
  in robots, 35
Speech Understanding Research (SUR) project, 30, 152
speed
  of hardware, 51, 98
  of machine translation of natural language, 27
  in research and development systems, 103
  of robot operations, 34
SRL (Schema Representation Language), 139
Stanford University, 78, 162, 163
state-space representations, 56-62
Steele, Guy L., Jr., 81, 142, 143

Strategic Computing Program, 153
STRIPS (robot teaching system), 35-36
STUDENT (natural language program), 162
subprocedures, 71
subroutines, in augmented transition networks, 70
supercomputers, 104
Superspeed Computer Project (Japan), 150
SUMEX-AIM (Stanford University Medical Experimental Computer Project), 165
Sussman, Gerald, 78
symbolic processing, 51-52, 77
  LISP for, 79
Symbolics, Inc., 105, 140
syntax, 69
syntax, of natural language, 23
SZKI, 87, 167

task decomposition, 42
teaching, of robots, 35-36
technical management, in development of expert systems, 130
TEIRESIAS (system), 166
Teknowledge, Inc., 138, 140
testing
  of expert systems, 133-34
  of XCON, 119-21
text critiquing, 30
theorem-proving based synthesis, 42
three-dimensional vision systems, 37, 39
TIMM (The Intelligent Machine Model; software tool), 138
top-down processing (backward chaining), 70
TOPS-20 operating system, 107-8
touch sensing, 39-40
training
  commercially available, 139-40
  from Digital Equipment Corporation's Educational Services, 141-43
  of engineers, in artificial intelligence, 131
transformation rules, 42
translation of natural language, 27-28
  conceptual dependencies in, 67
trees (graphs), 57
Turing, Alan, 11, 155, 161

Turing Institute (United Kingdom), 155-56, 168
TVX (The VMS Expert; expert system), 124
two-dimensional vision systems, 37, 39

ULTRIX operating system, 106
Unimation, Inc., 79
United Kingdom, 154-56
University of Southern California Information Sciences Institute (ISI), 44, 136
UNIX operating system, 106
USE.IT (automatic programming system), 43
users
  delivery of expert systems to, 134-35
  in design of expert systems, 53, 129, 133
  of XCON, 117-18

V (language), 43
VAL (language), 79
values, inheritance of, 68-69
Vance, Cyrus, 29
variables, 52
  in logic, 62
VAX 8600 systems, 106
VAX LISP (language), 73-74, 82, 136
  training in, 142
VAX OPS5 (language), 73-74, 136
VAXstation II, 107
VAX systems, 105-7
  XCON expert system for configuration of, 112-13
virtual memory, 105
vision systems, 36-39
VMS operating system, 105, 106
  expert systems developed on, 112
  TVX expert system for, 124
vocabularies
  of machine translation systems, 27
  of speech recognition systems, 30
  see also dictionaries
von Neumann architecture, 148

Waltz, David L., 58, 166
Warren, David H.D., 83-84
Waters, Richard C., 44
Weidner Communications Corporation, 28
Weizenbaum, Joseph, 163
West Germany, 156
Wilks, Yorick, 67
windows, in LISP machines, 104
Winograd, Terry, 78, 164
Winston, Patrick H., 8, 108, 163, 166
Woods, William, 70, 165
word processing, 77
word size, 97-98
working memory, in OPS5, 88
workstations, 101
  as delivery systems, 102
  for research and development, 104-5
  VAX-based, 107
Wright, J.M., 139

XCON (expert system), 19-20, 56, 89, 167
  development of, 112-17, 124-26
  integration into organization of, 121
  performance and impact of, 122-24
  testing of, 119-21
  use of, 117-19
Xerox 1100 (LISP machine), 105
Xerox Palo Alto Research Center, 82, 91, 105
XSEL (expert SELling tool; expert system), 117, 124
XSITE (eXpert SITE preparation tool; expert system), 124

Yale University, 25, 28-29

ZETALISP (language), 82, 83, 91